Name _____ Date _____

Find the missing numbers. Solid lines mean multiply.
Dotted lines mean add.

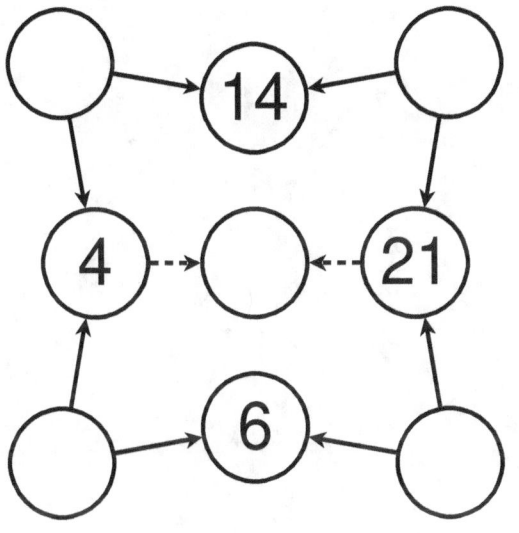

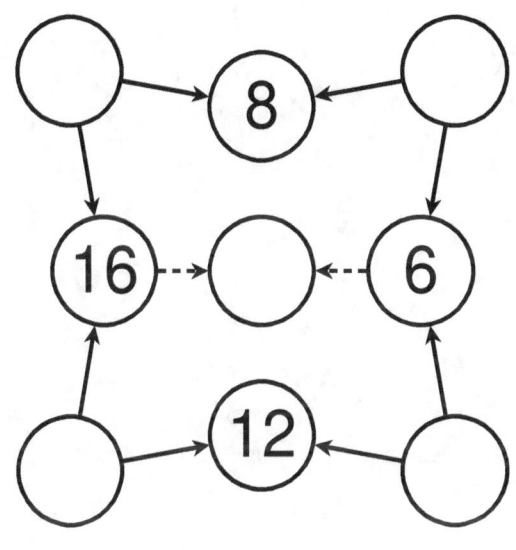

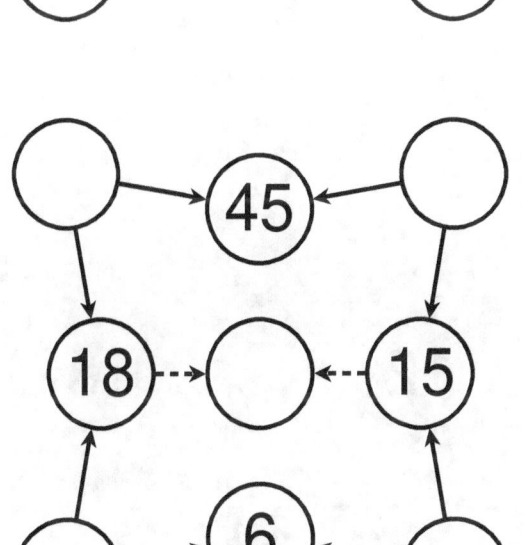

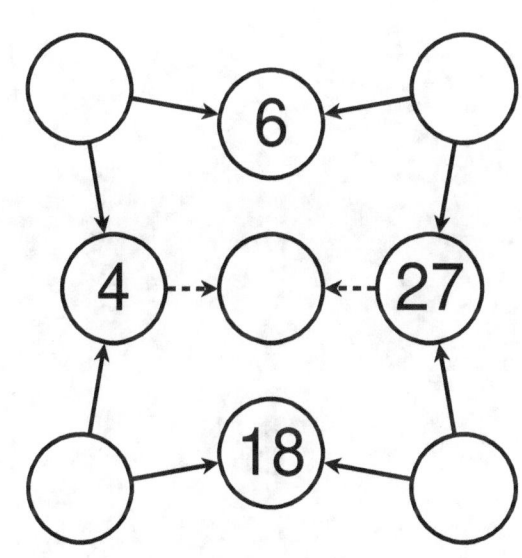

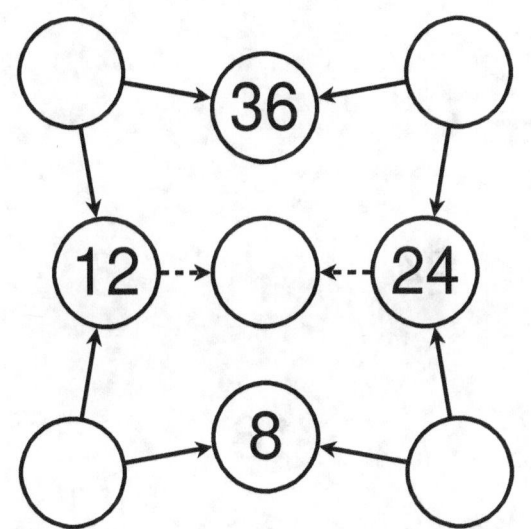

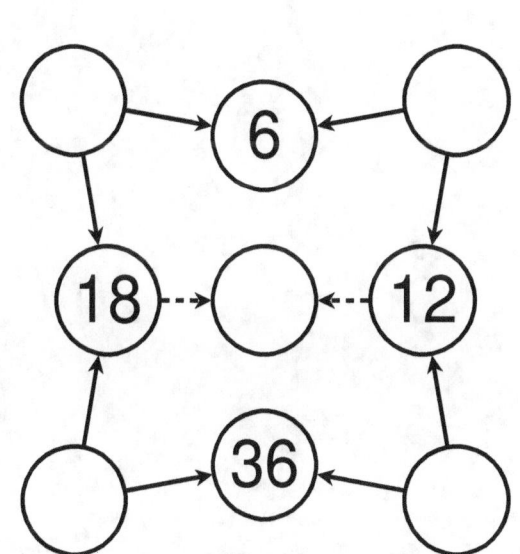

Name _____ Date _____

Find the missing numbers. Solid lines mean multiply.
Dotted lines mean add.

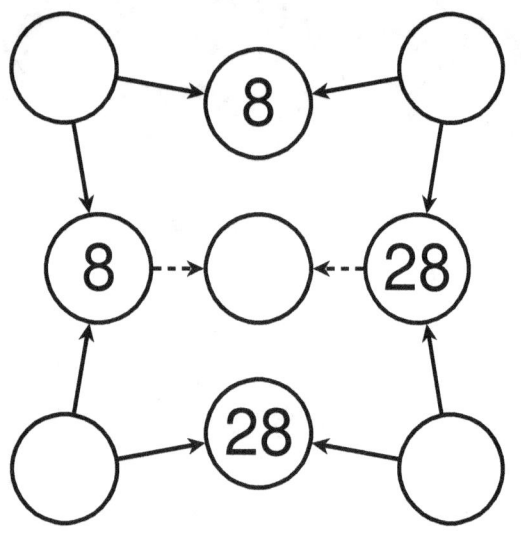

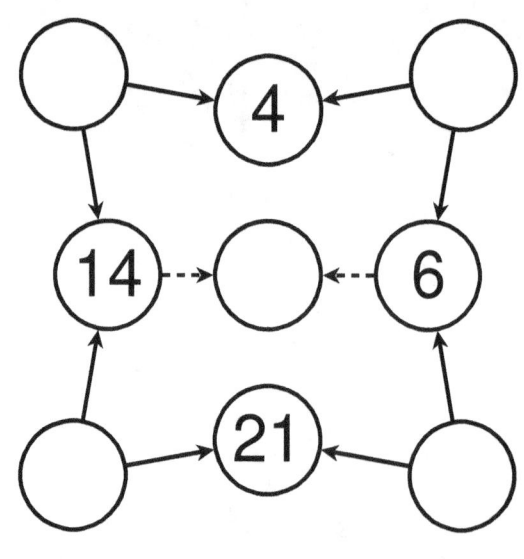

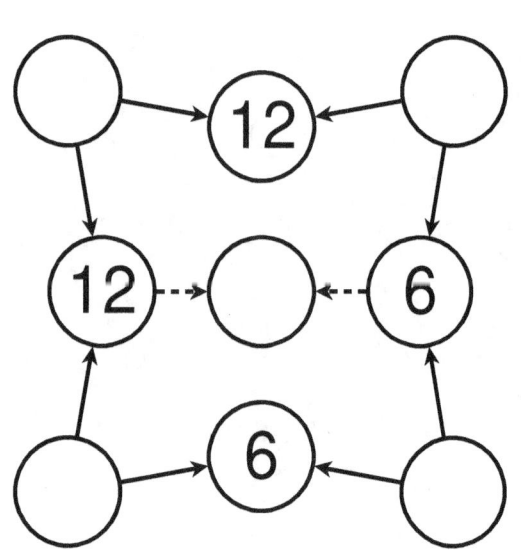

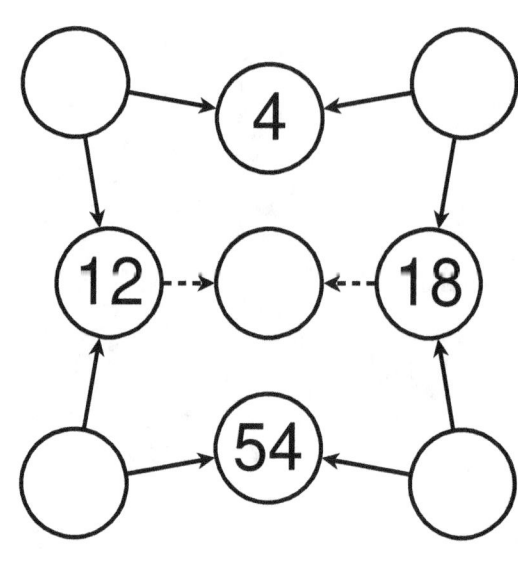

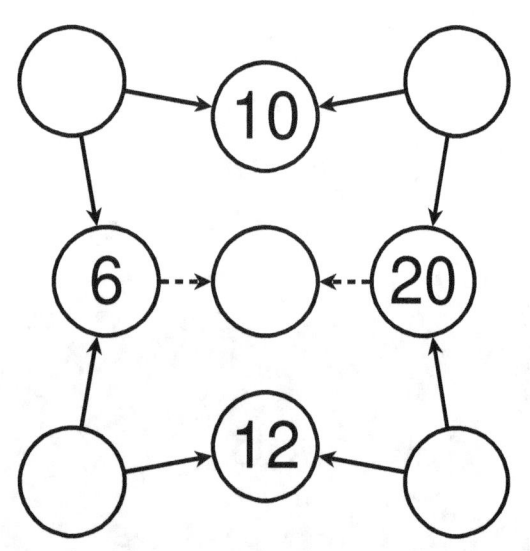

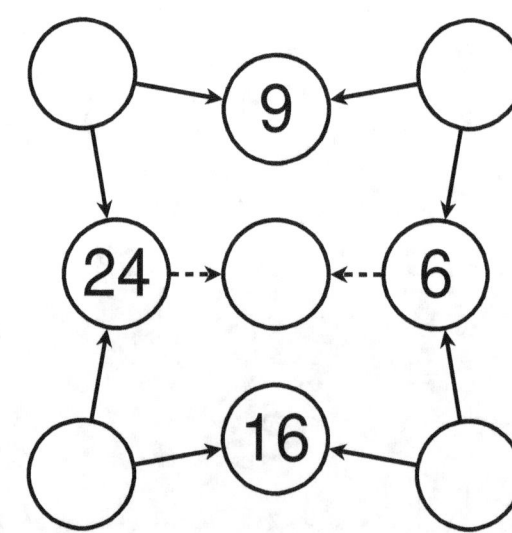

Name _____ Date _____

Find the missing numbers. Solid lines mean multiply.
Dotted lines mean add.

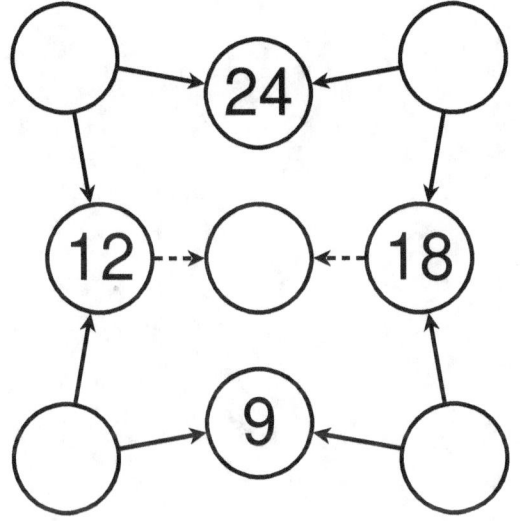

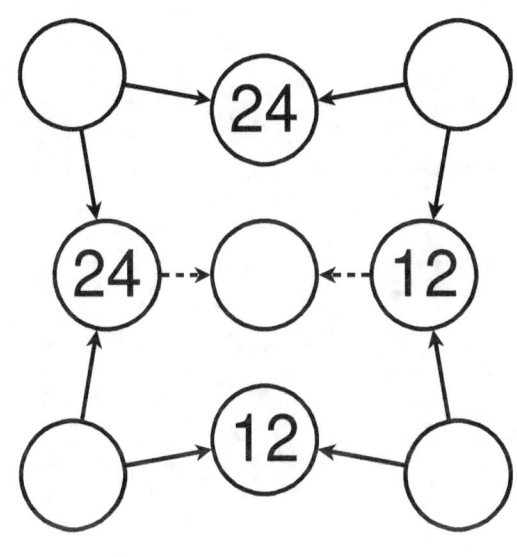

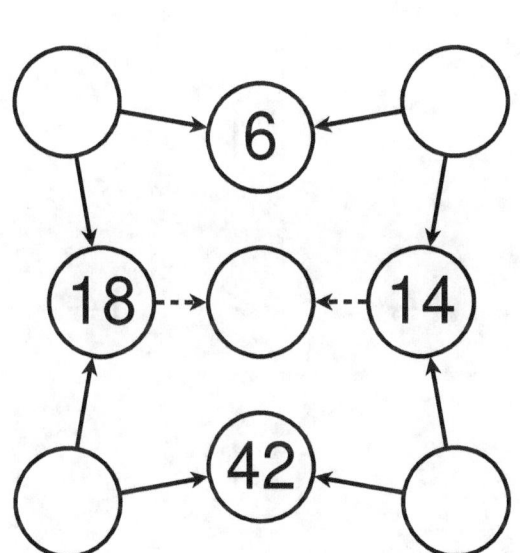

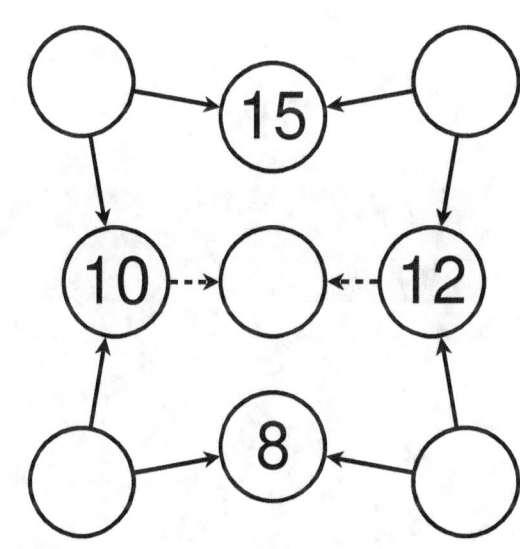

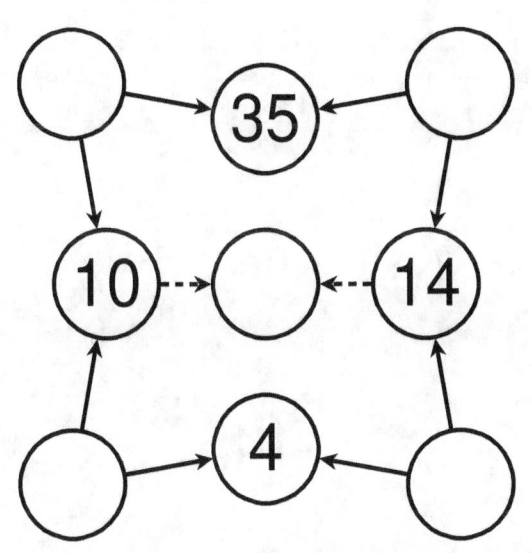

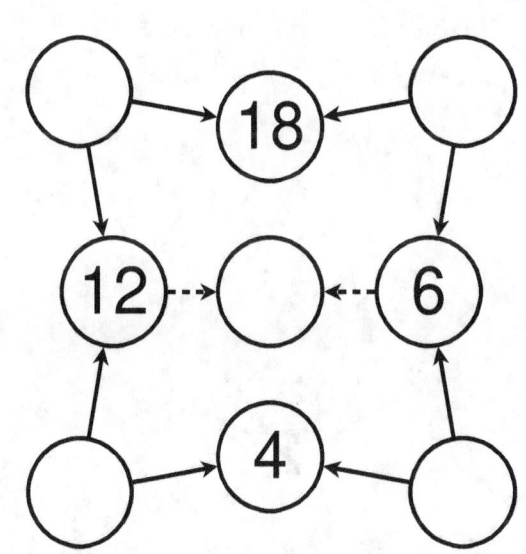

Name _____ Date _____

Find the missing numbers. Solid lines mean multiply.
Dotted lines mean add.

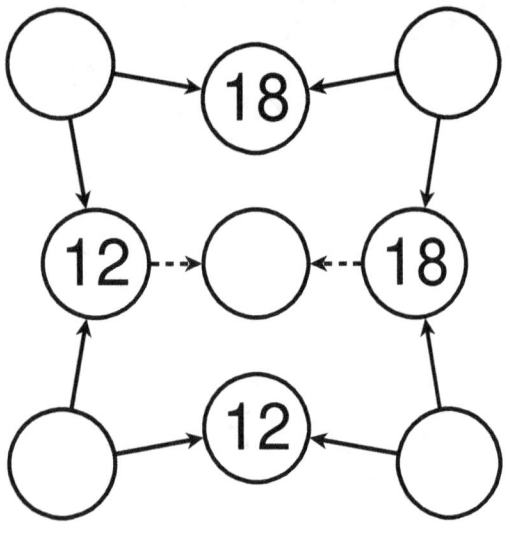

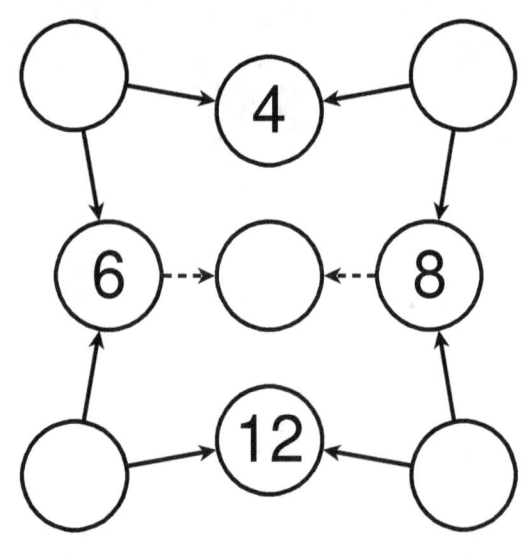

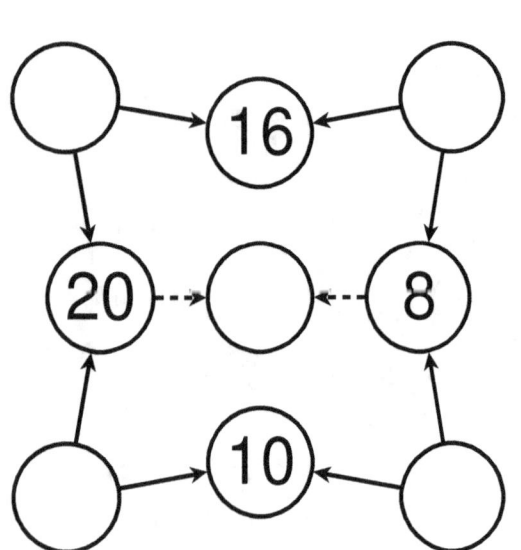

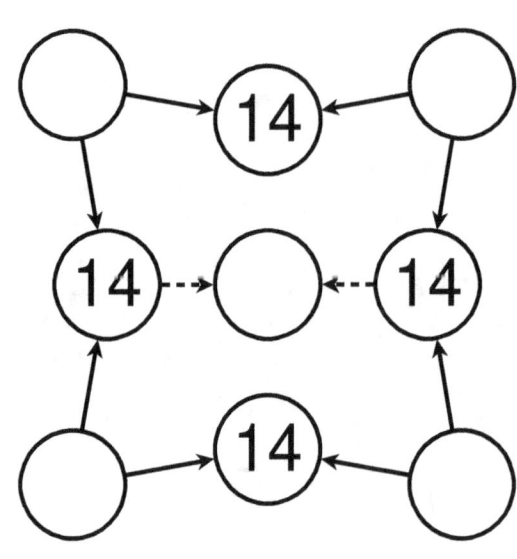

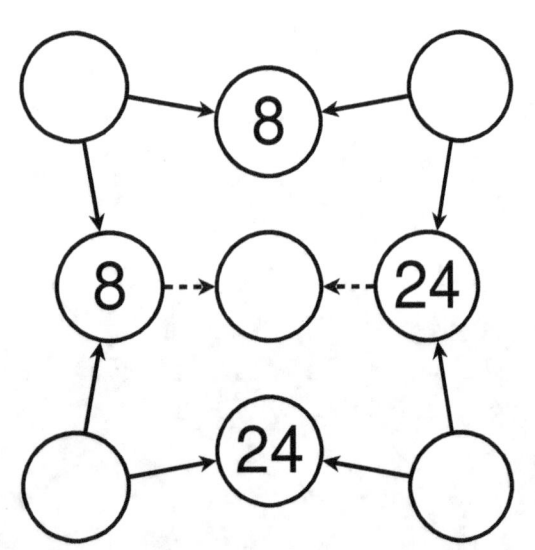

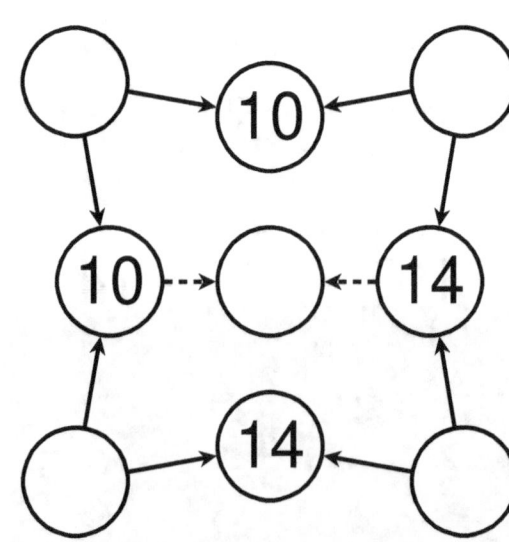

Name _____ Date _____

Find the missing numbers. Solid lines mean multiply.
Dotted lines mean add.

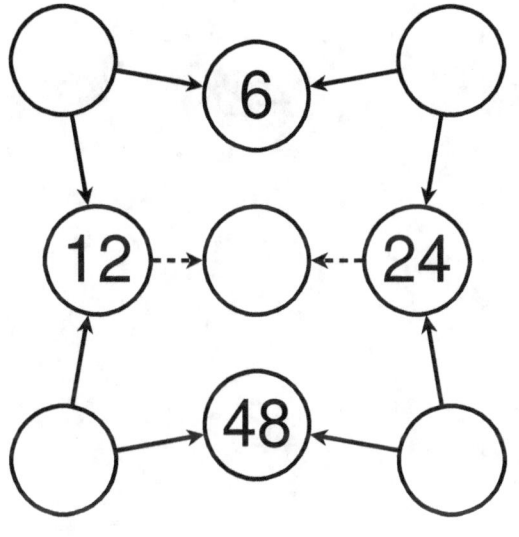

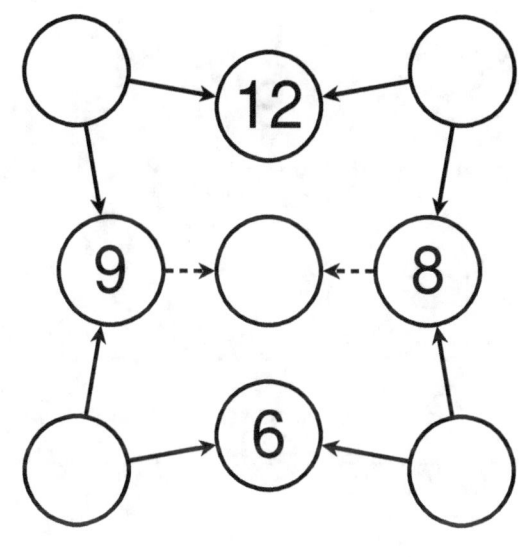

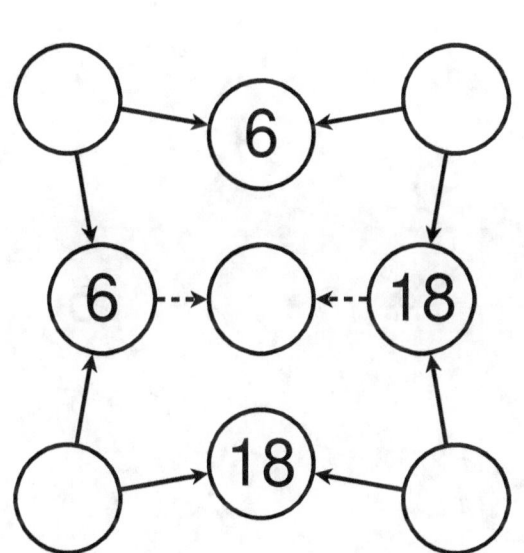

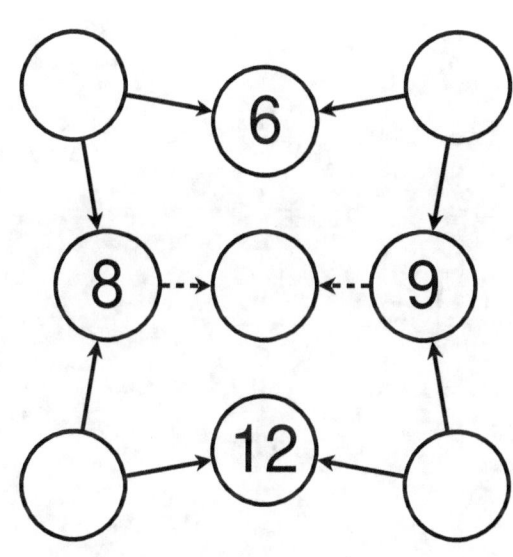

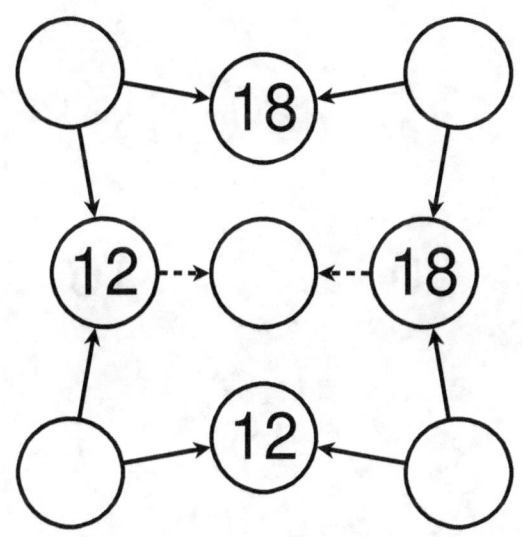

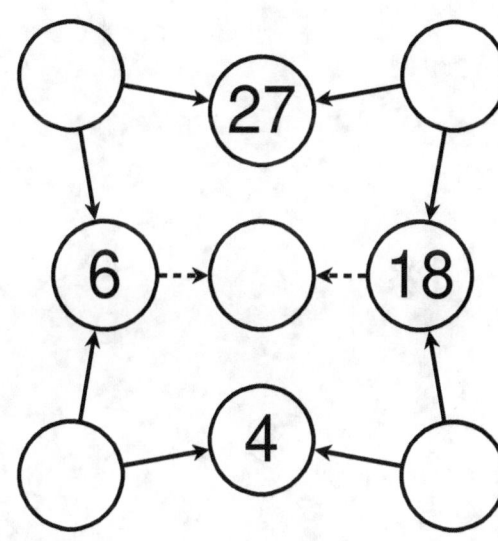

Name _____ Date _____

Find the missing numbers. Solid lines mean multiply.
Dotted lines mean add.

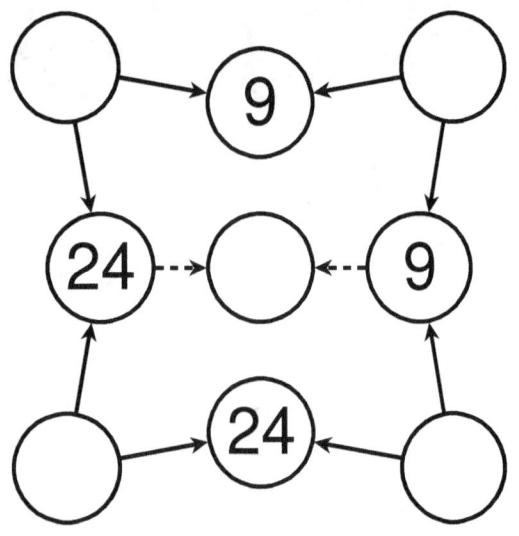

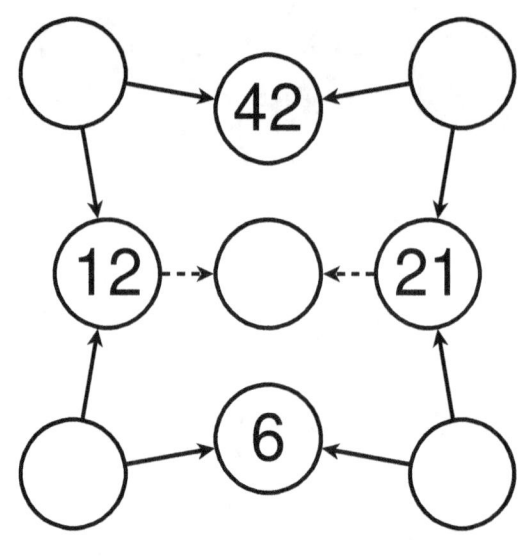

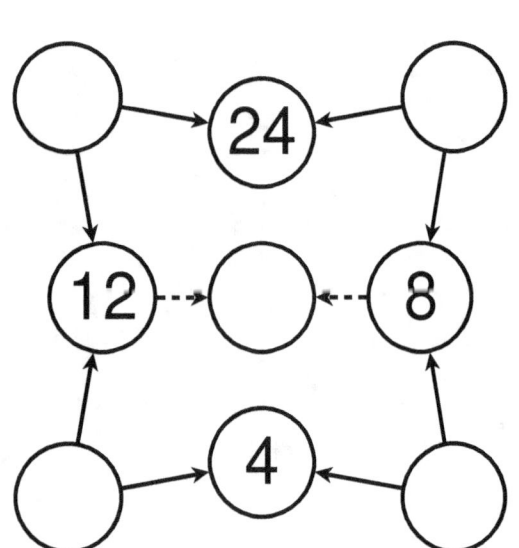

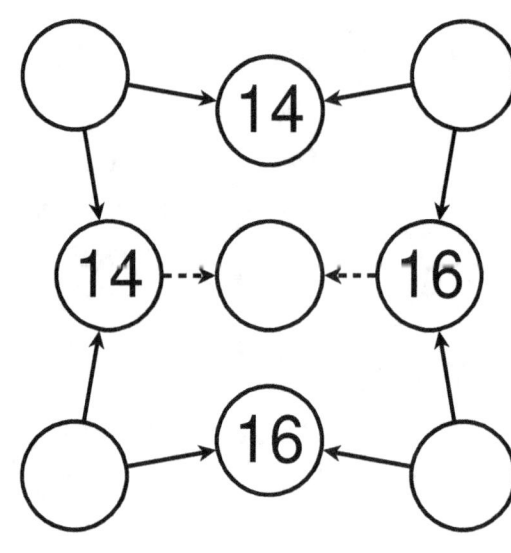

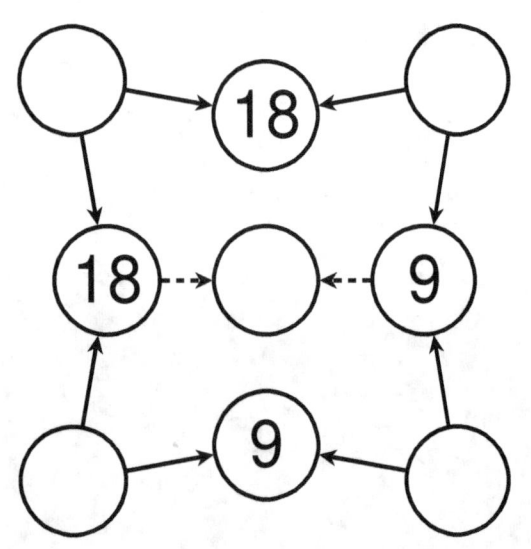

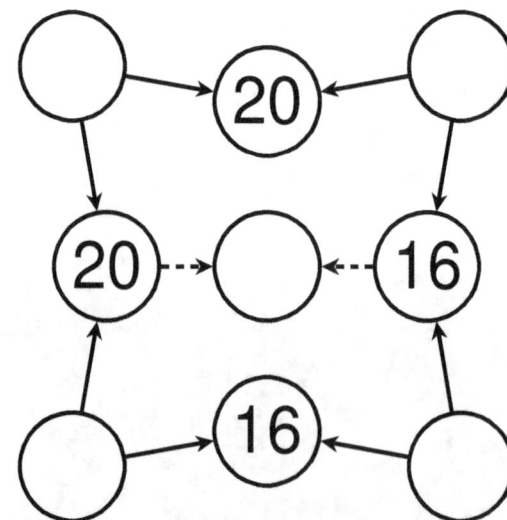

Name _____ Date _____

Find the missing numbers. Solid lines mean multiply.
Dotted lines mean add.

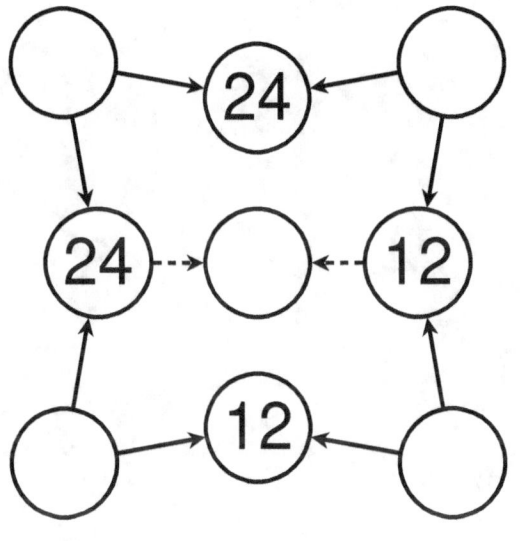

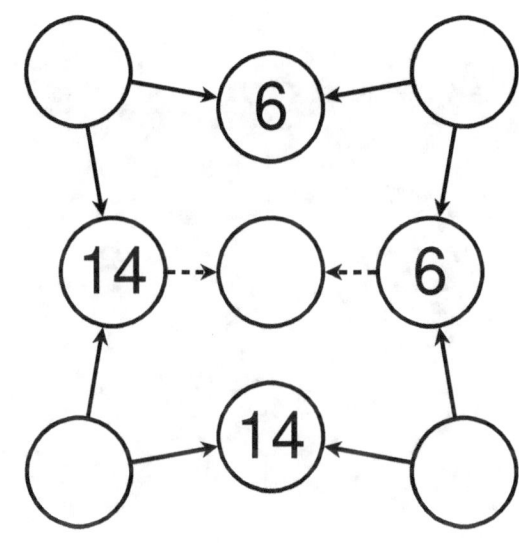

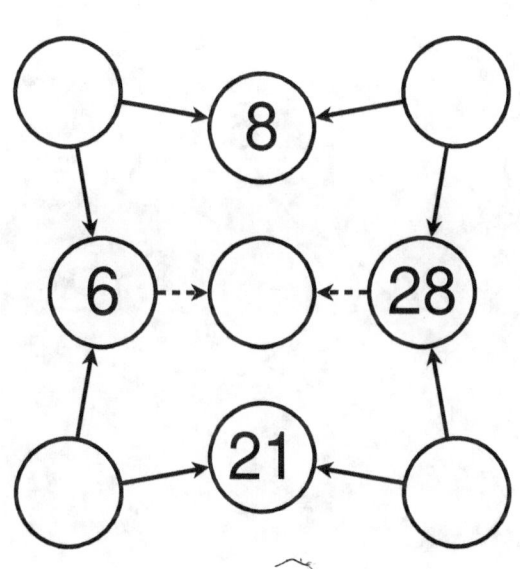

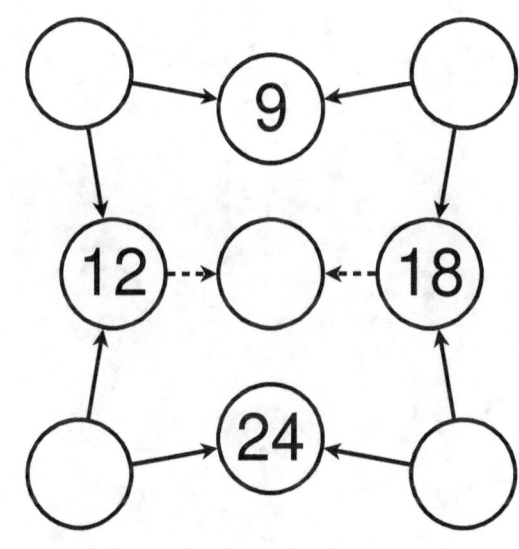

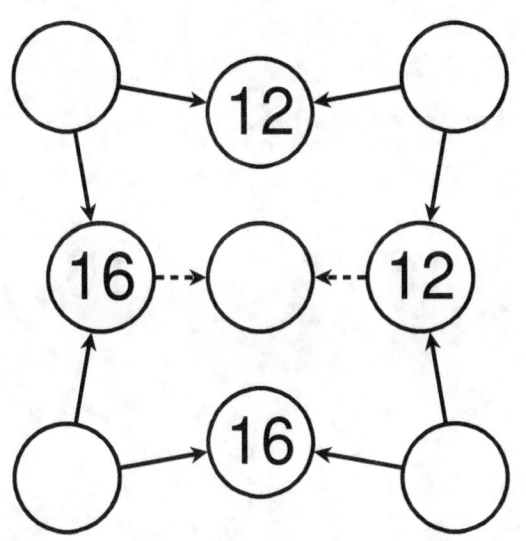

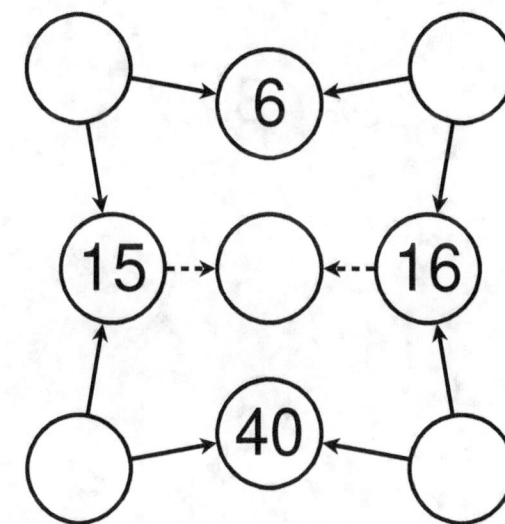

Name _____ Date _____

Find the missing numbers. Solid lines mean multiply.
Dotted lines mean add.

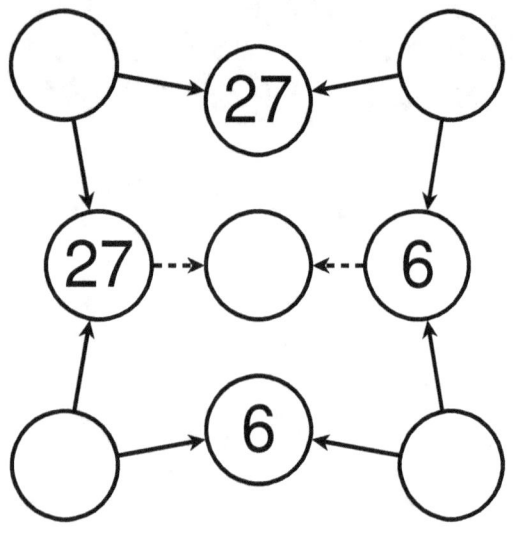

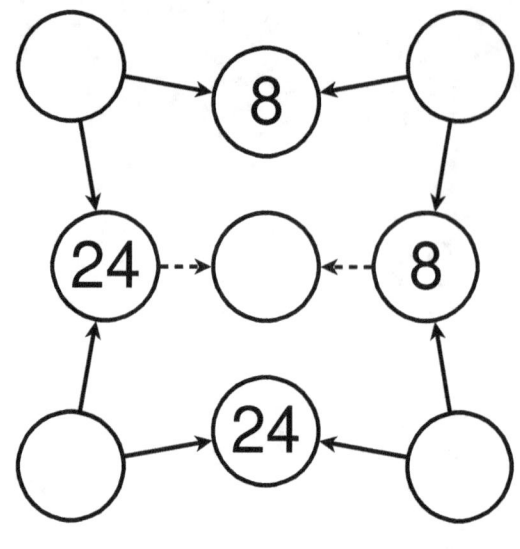

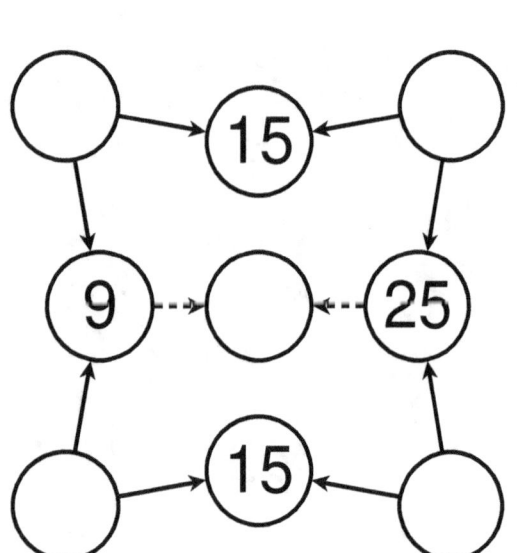

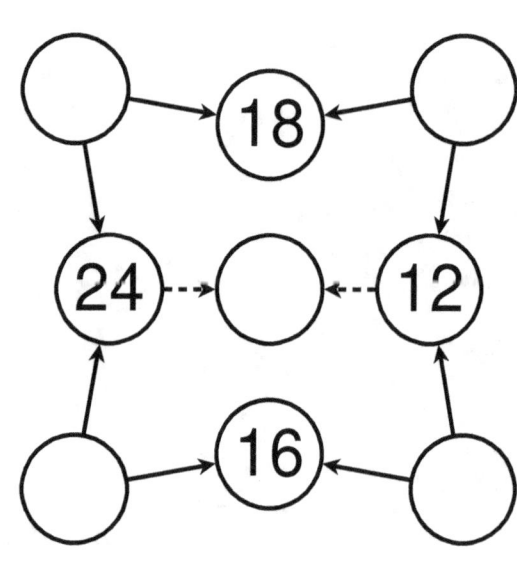

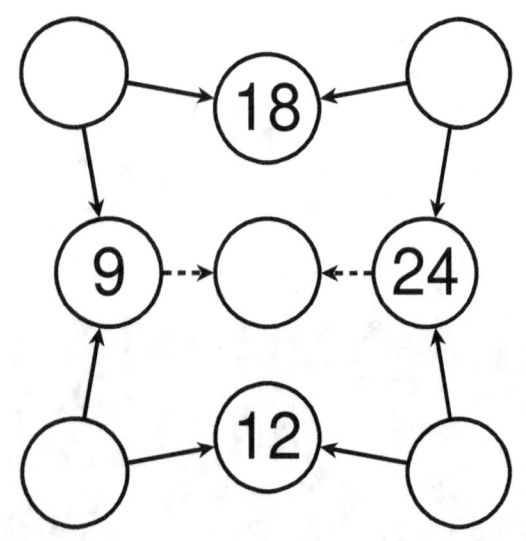

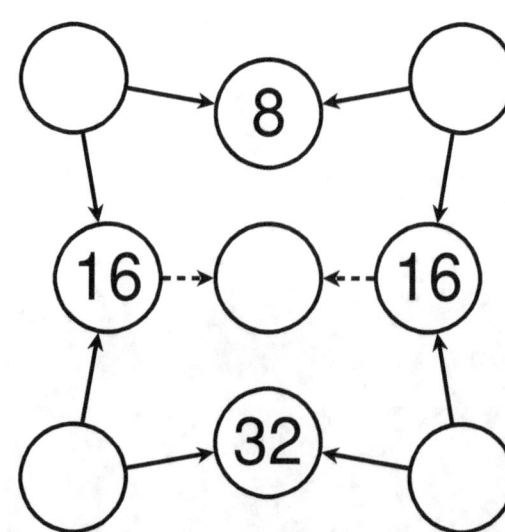

Find the missing numbers. Solid lines mean multiply.
Dotted lines mean add.

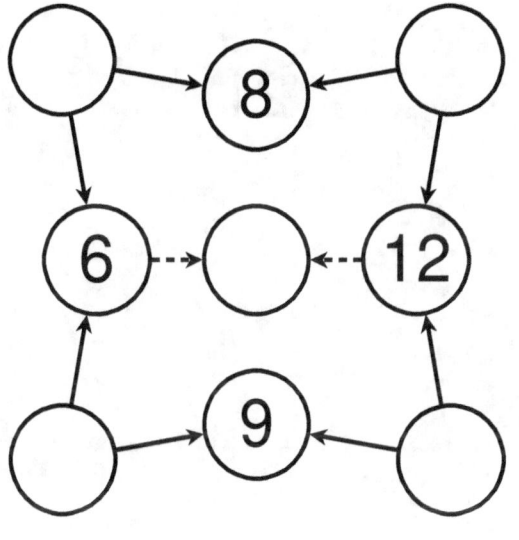

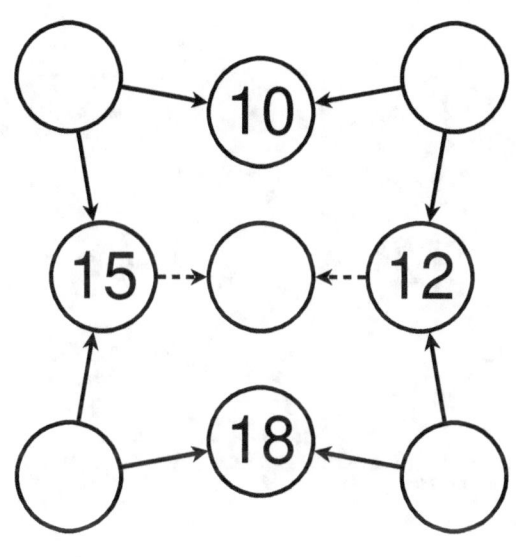

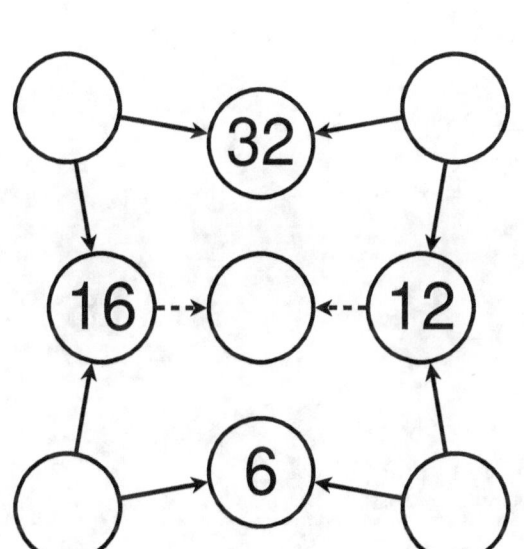

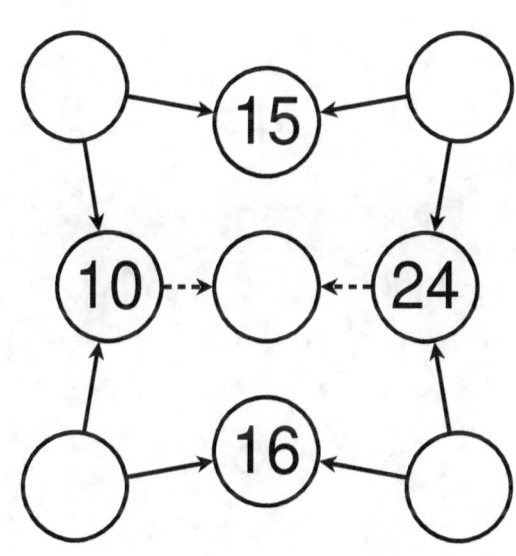

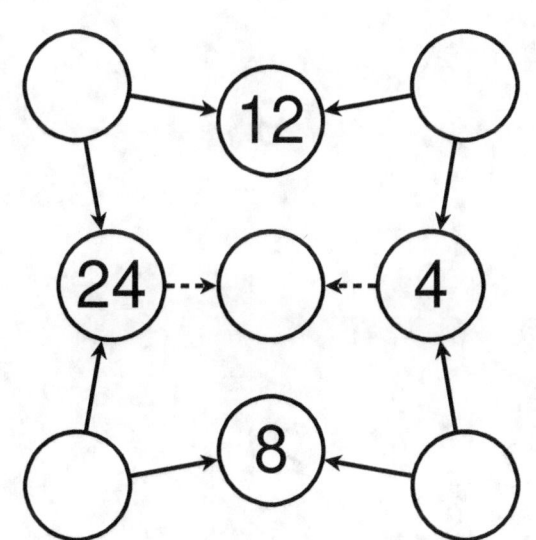

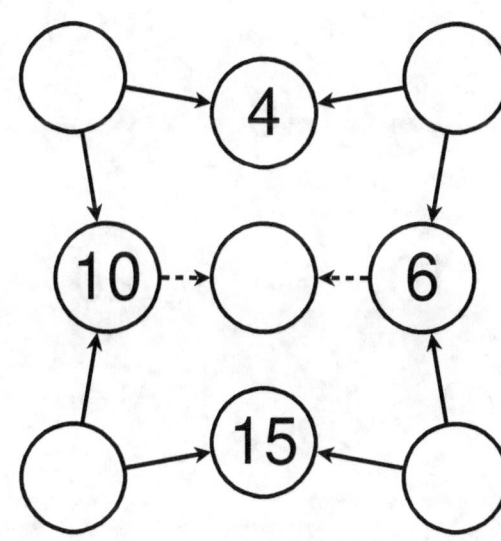

Name _____ Date _____

Find the missing numbers. Solid lines mean multiply.
Dotted lines mean add.

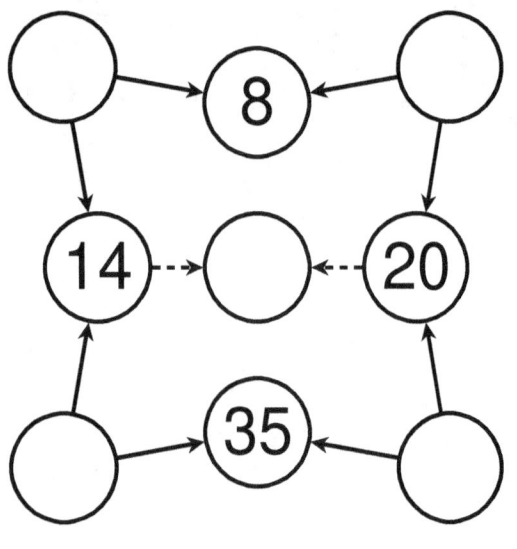

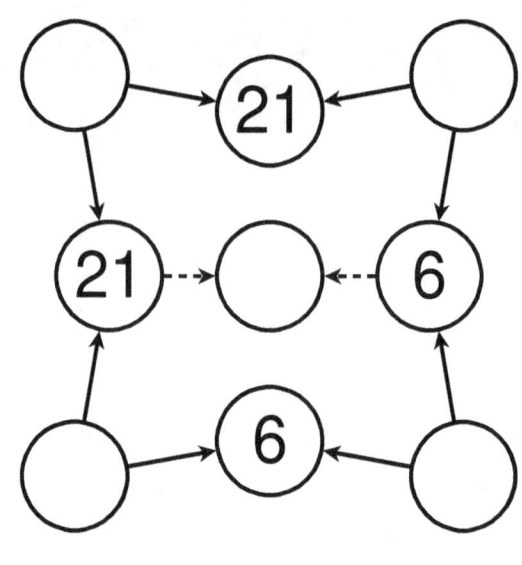

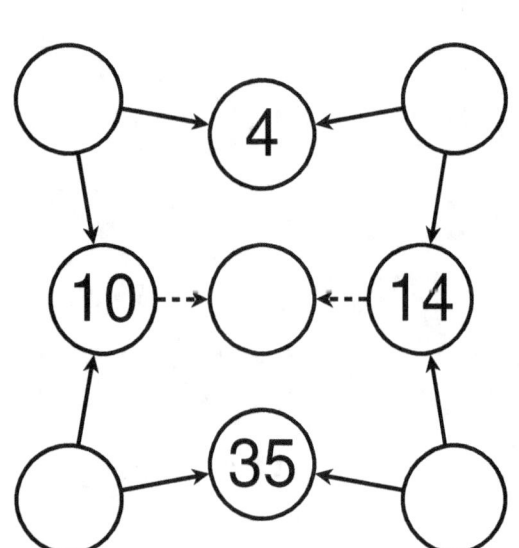

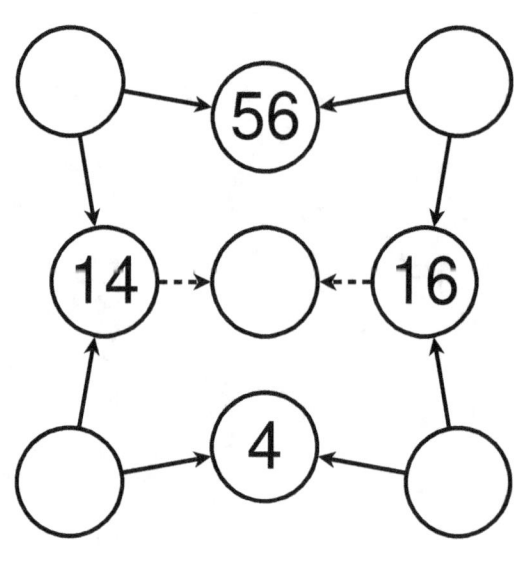

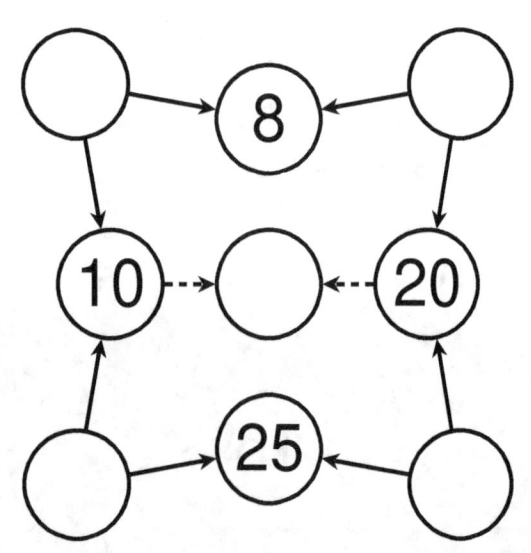

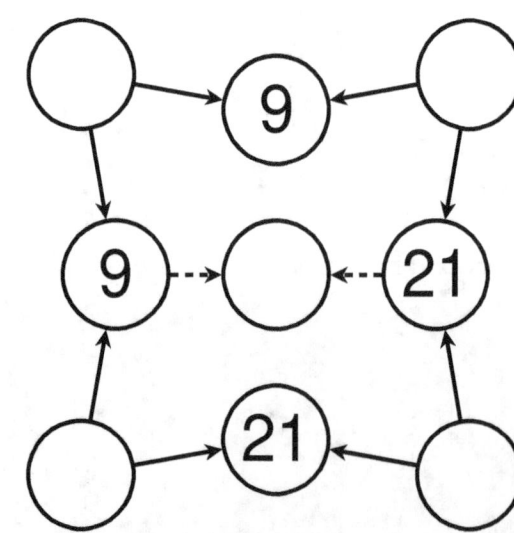

Page 10

Name _____   Date _____

Find the missing numbers. Solid lines mean multiply.
Dotted lines mean add.

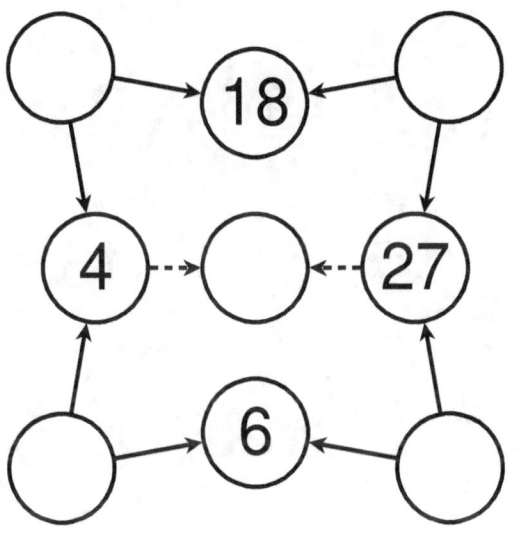

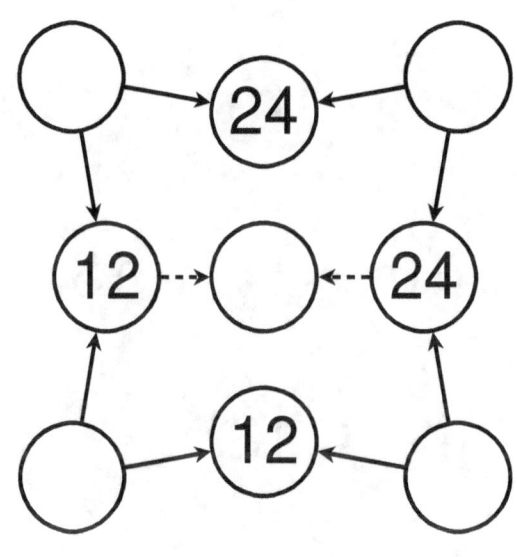

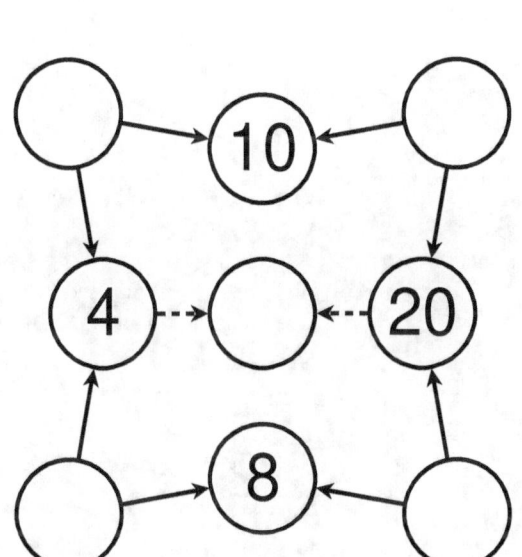

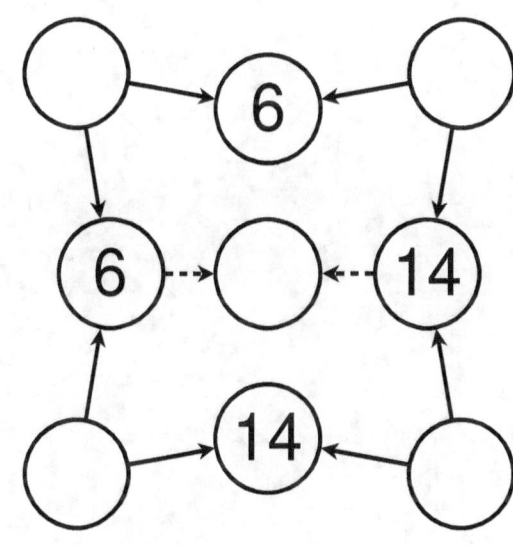

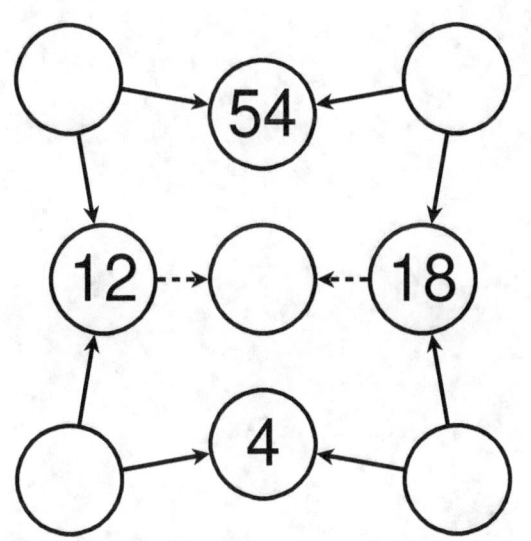

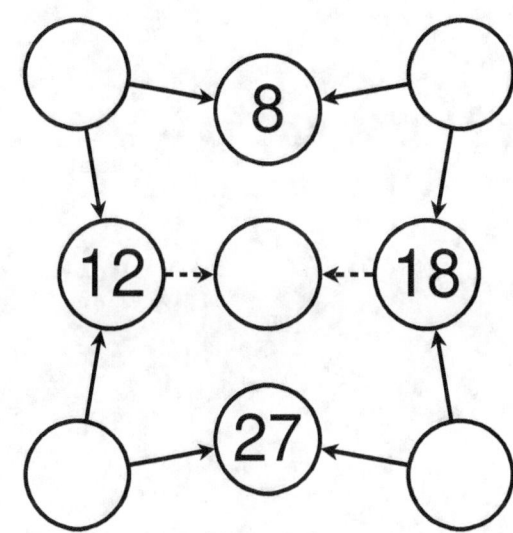

Name _____ Date _____

Find the missing numbers. Solid lines mean multiply.
Dotted lines mean add.

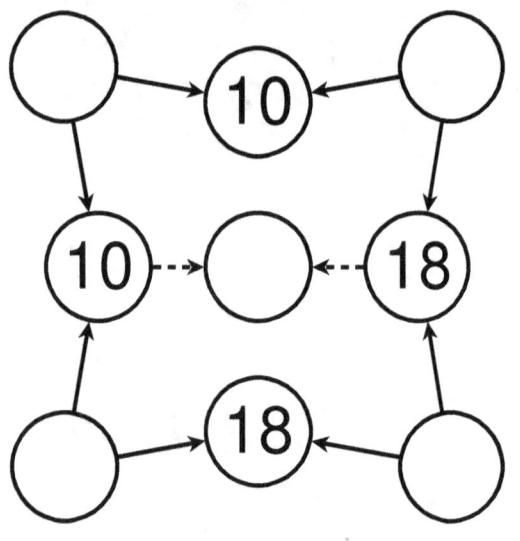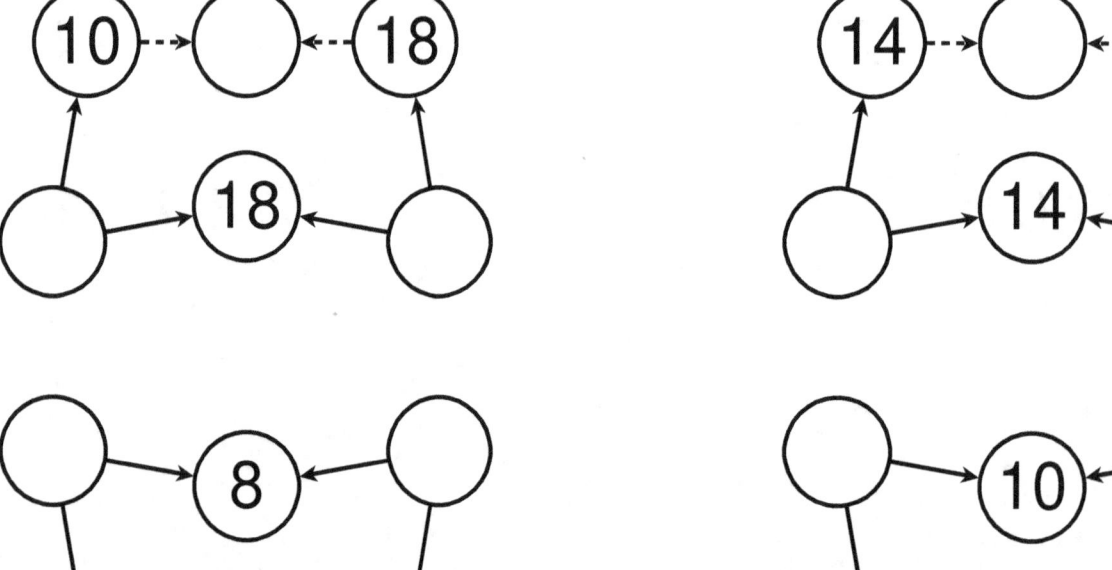

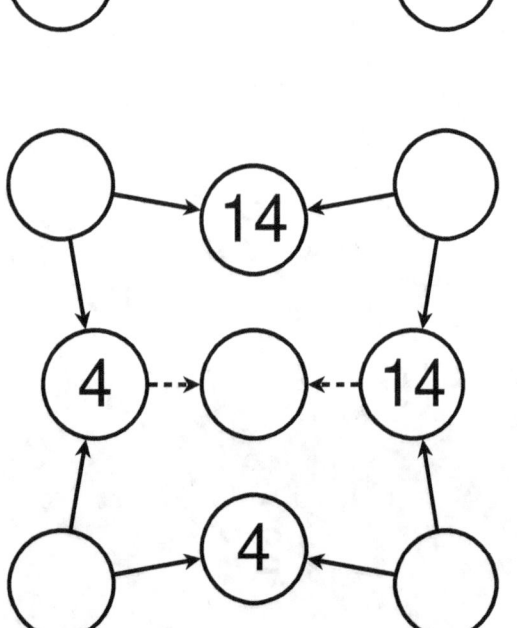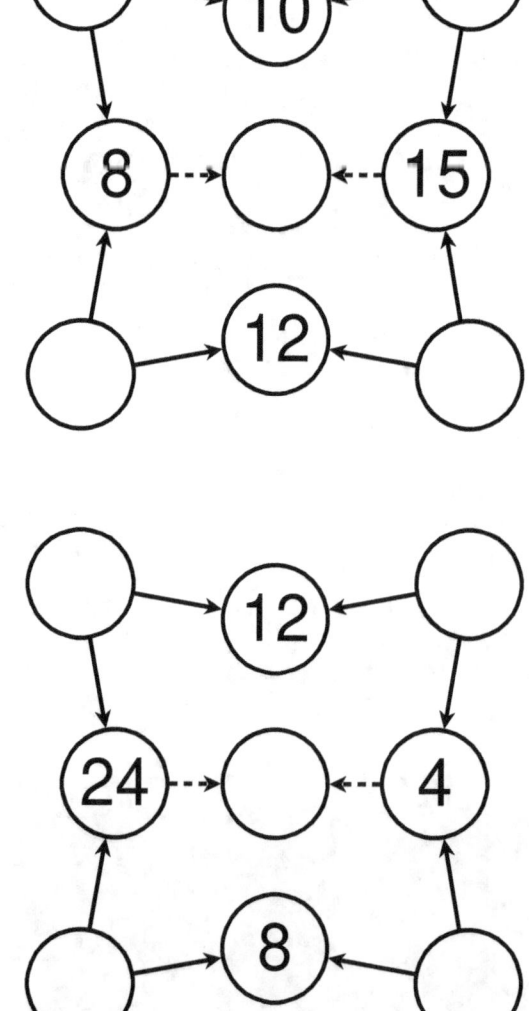

Name _____ Date _____

Find the missing numbers. Solid lines mean multiply.
Dotted lines mean add.

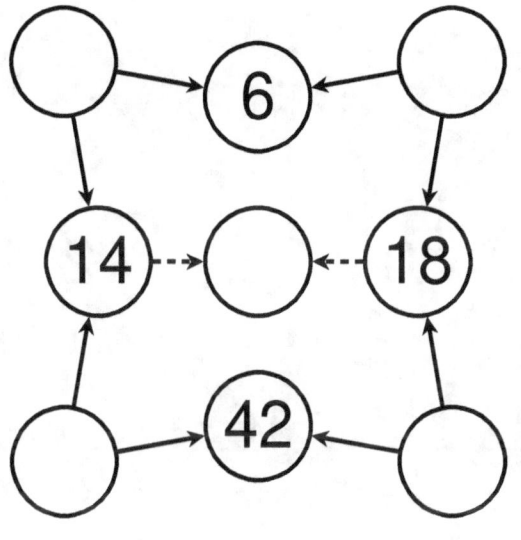

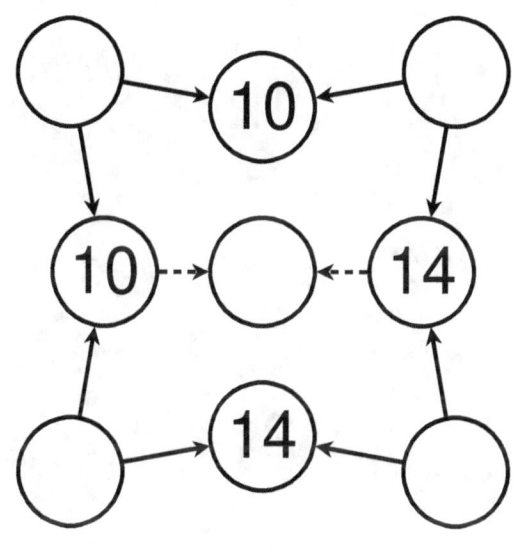

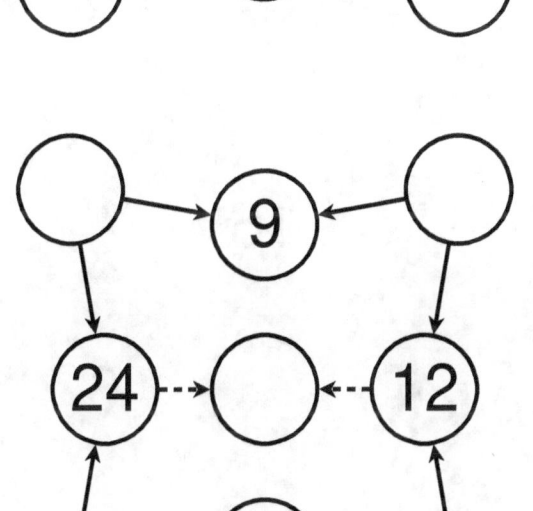

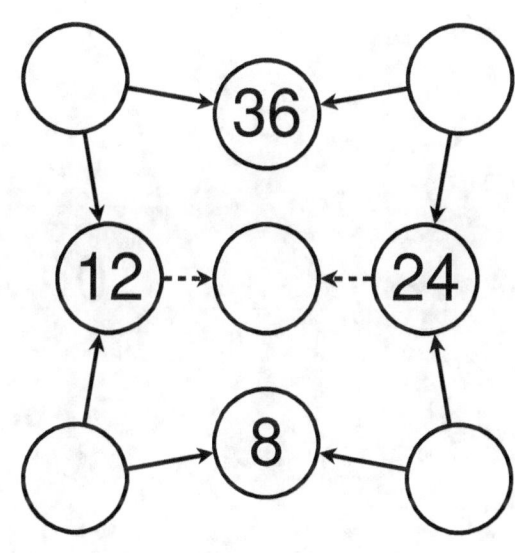

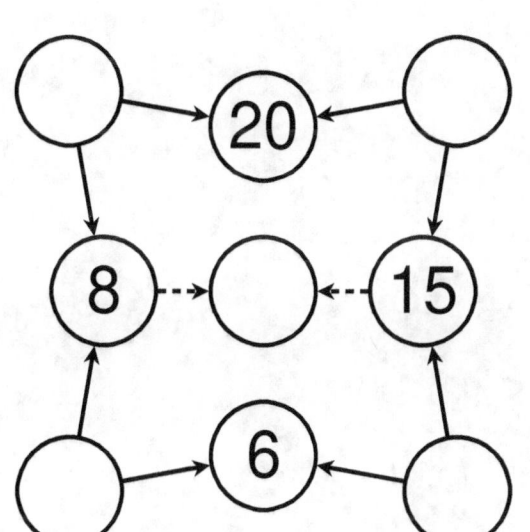

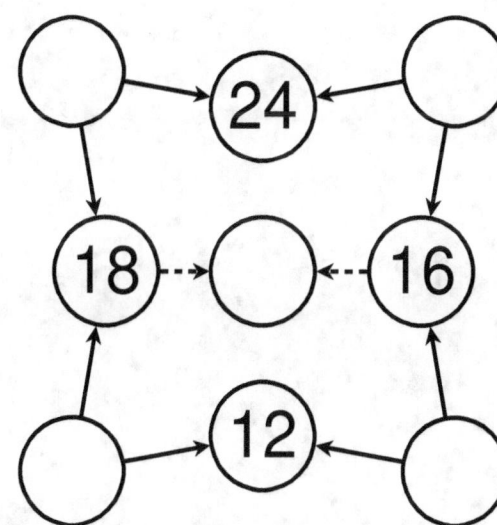

Name _____ Date _____

Find the missing numbers. Solid lines mean multiply.
Dotted lines mean add.

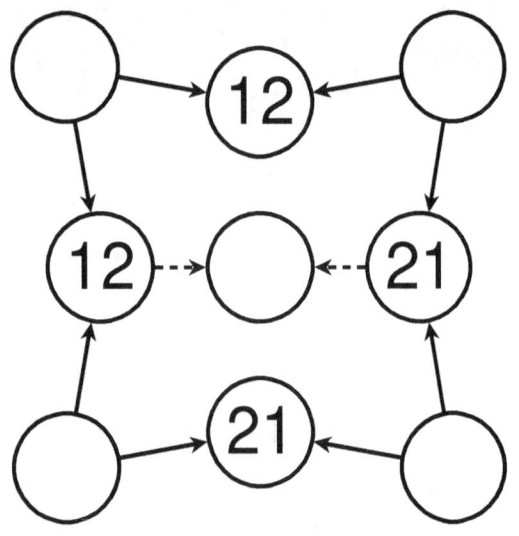

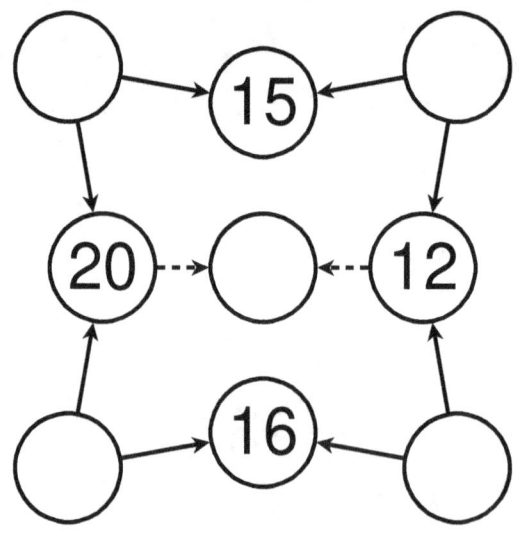

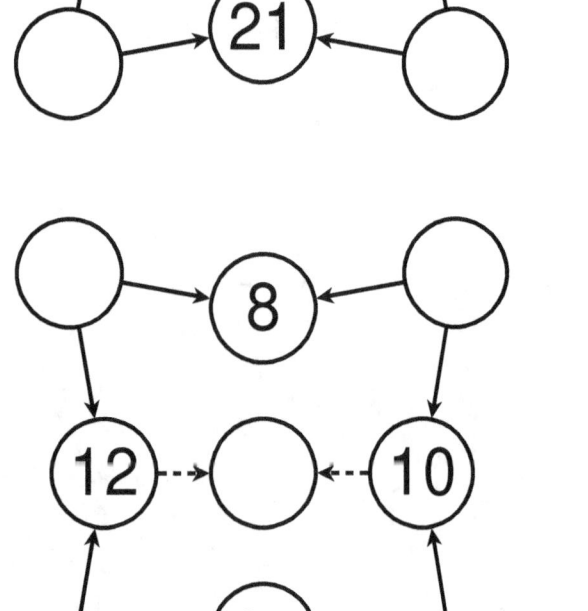

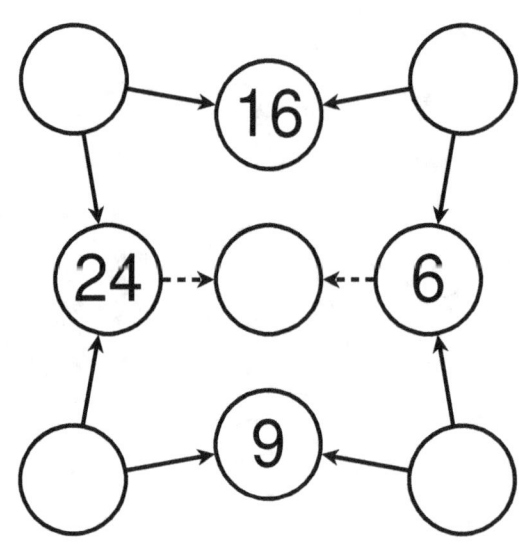

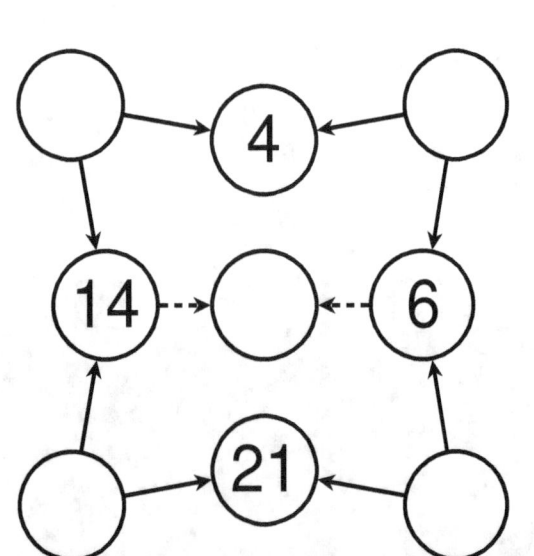

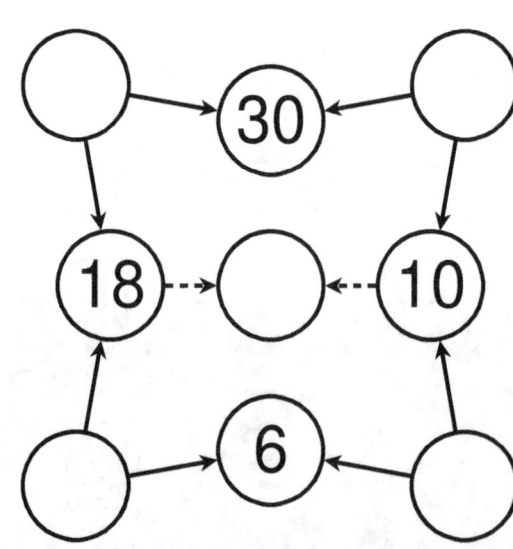

Name _____   Date _____

Find the missing numbers. Solid lines mean multiply.
Dotted lines mean add.

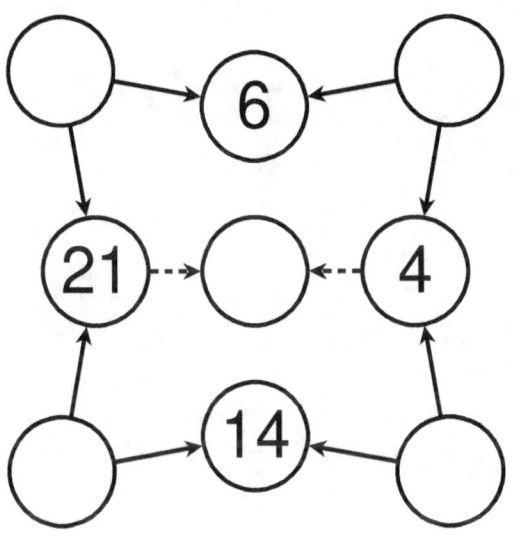

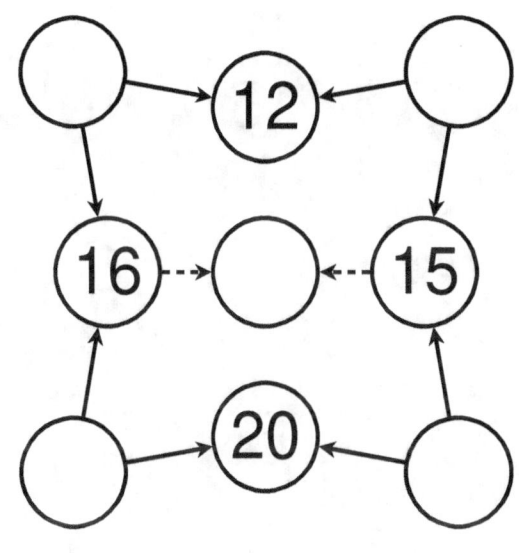

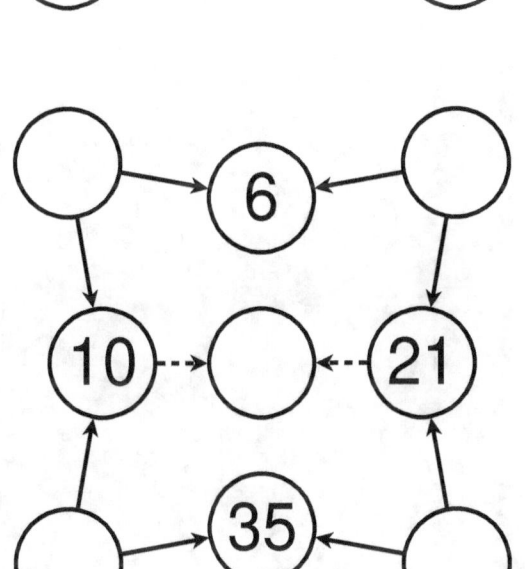

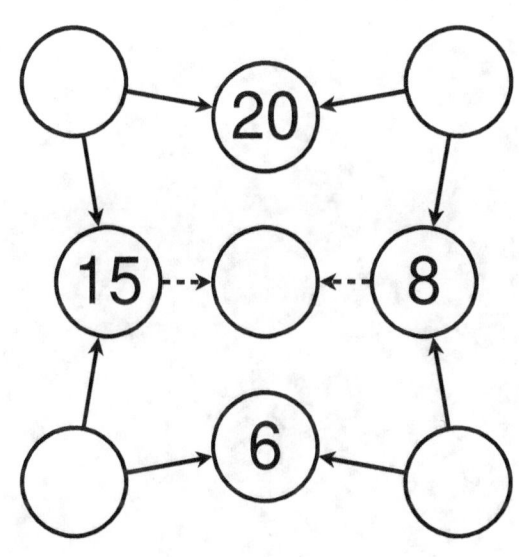

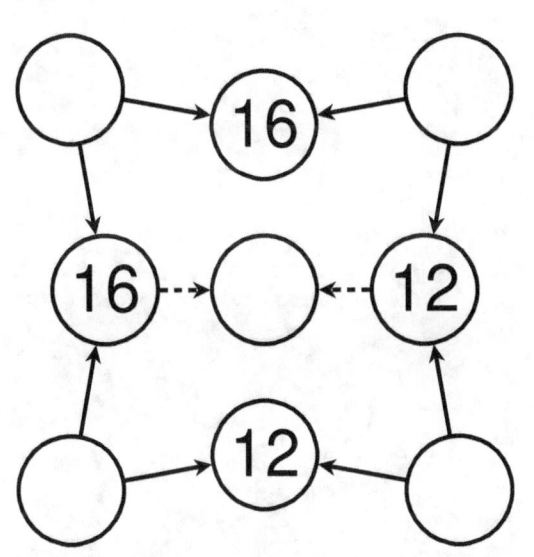

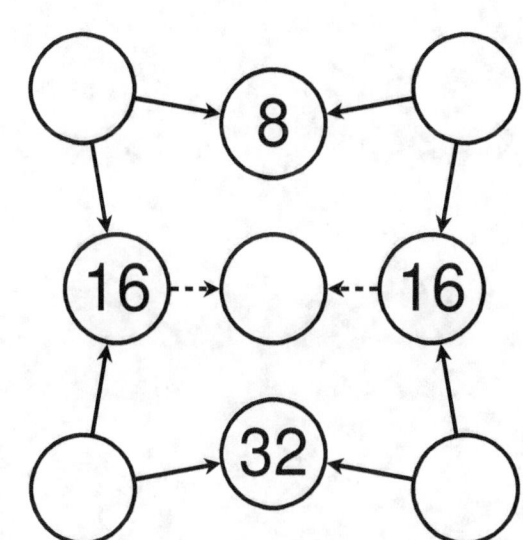

Name _____ Date _____

Find the missing numbers. Solid lines mean multiply.
Dotted lines mean add.

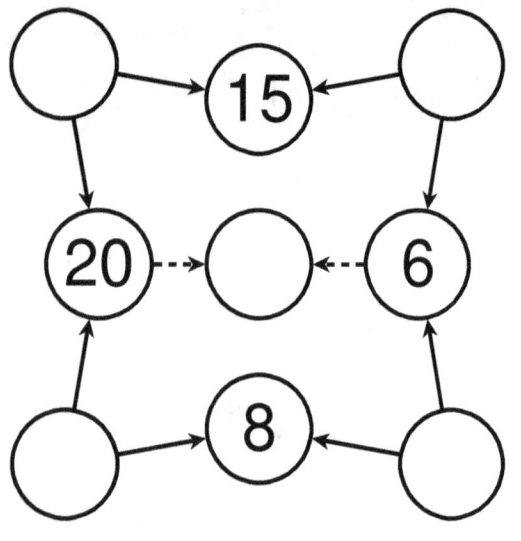

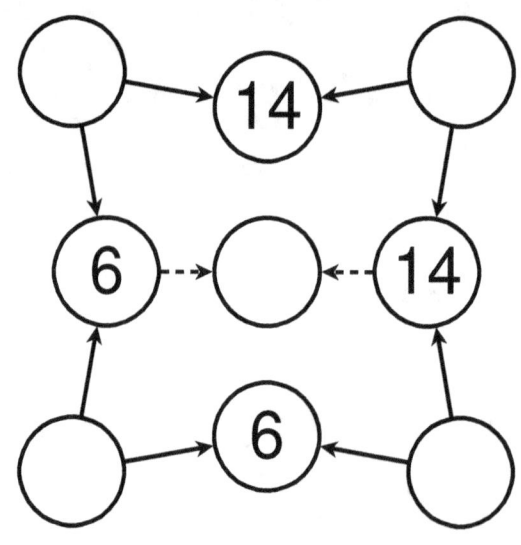

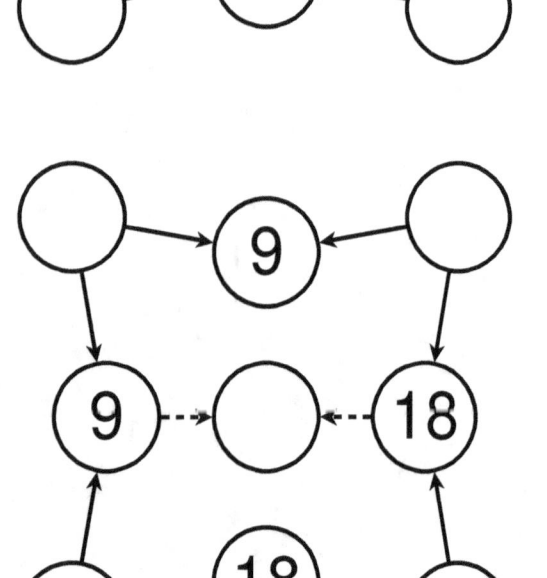

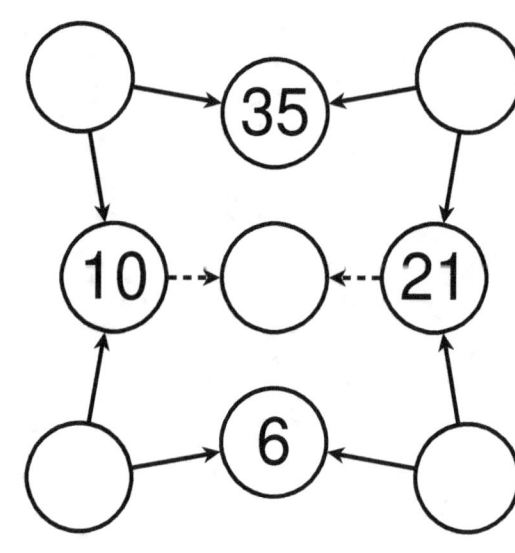

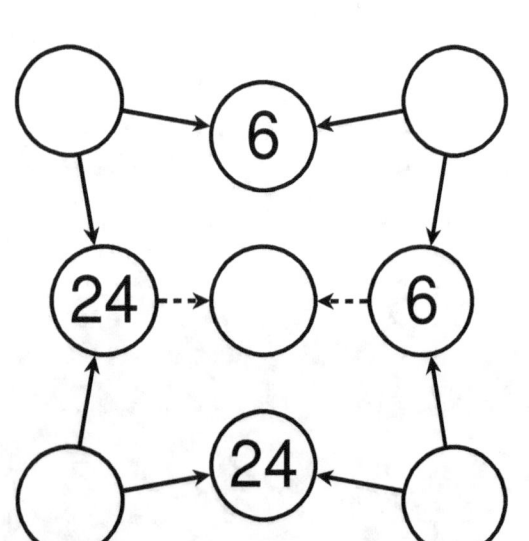

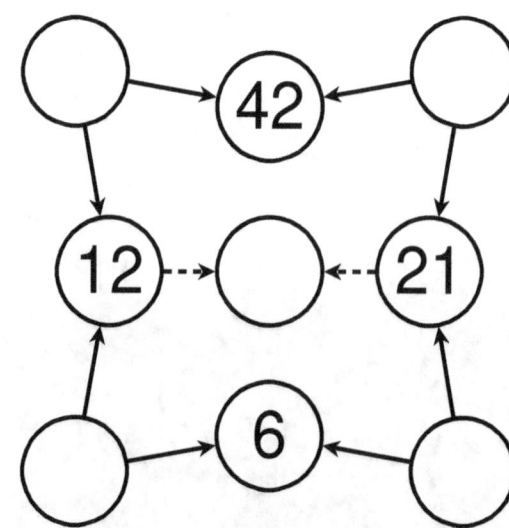

Name _____ Date _____

Find the missing numbers. Solid lines mean multiply.
Dotted lines mean add.

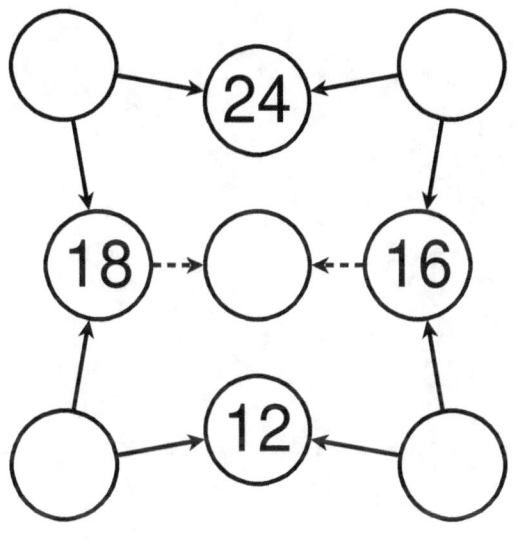

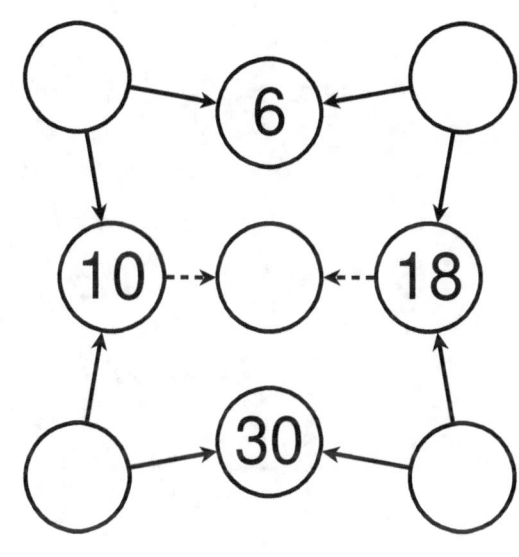

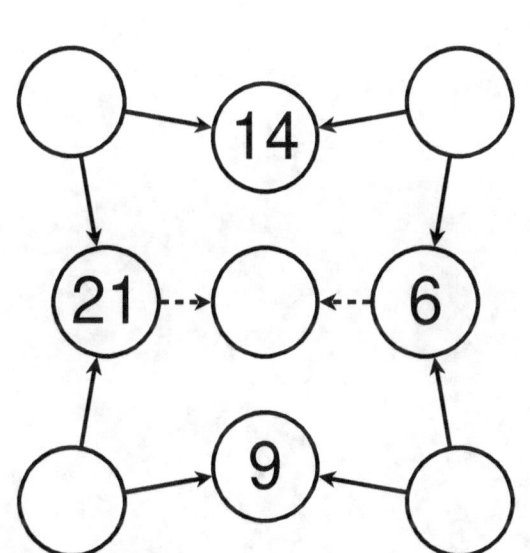

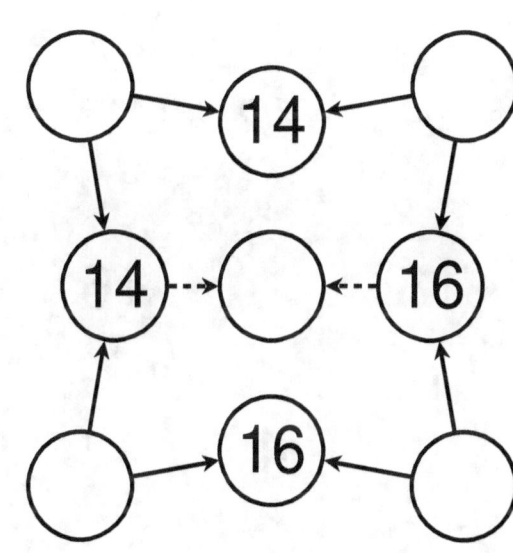

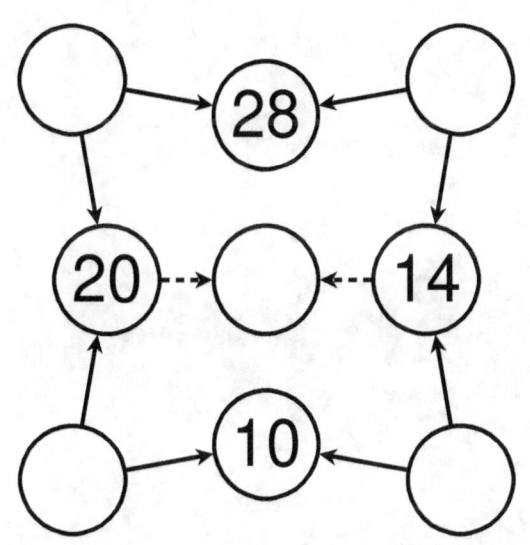

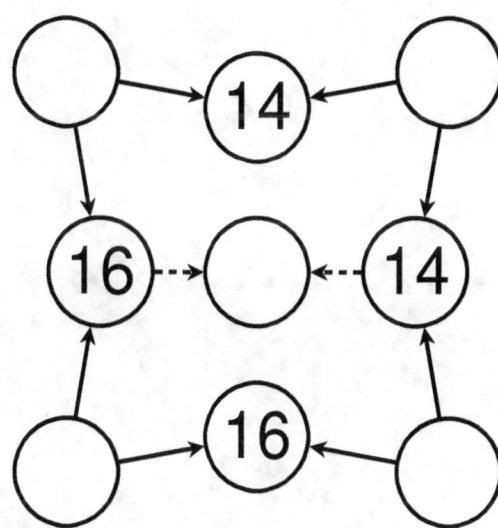

Name _____ Date _____

Find the missing numbers. Solid lines mean multiply.
Dotted lines mean add.

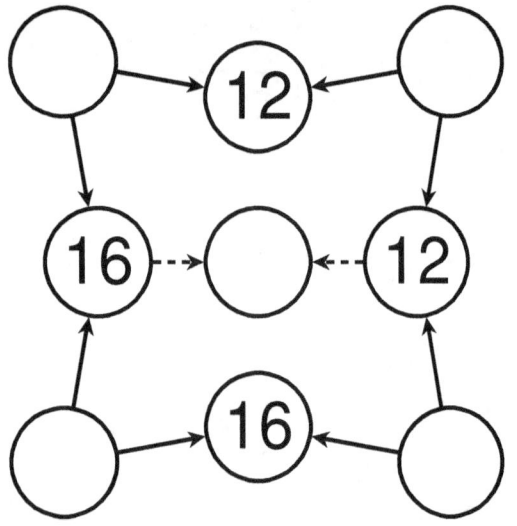

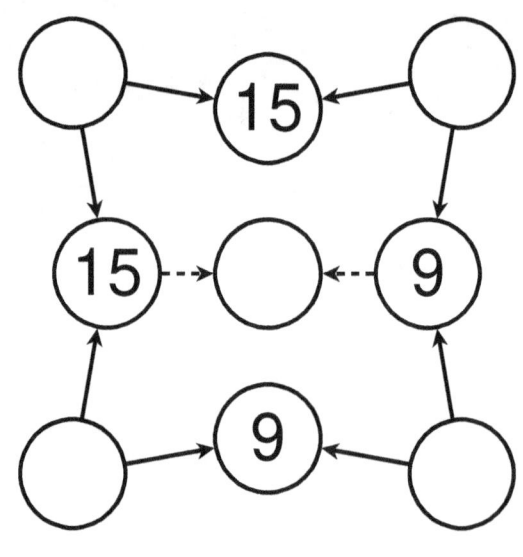

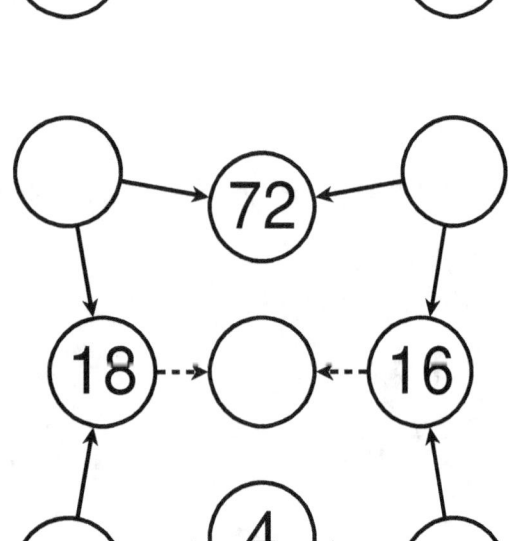

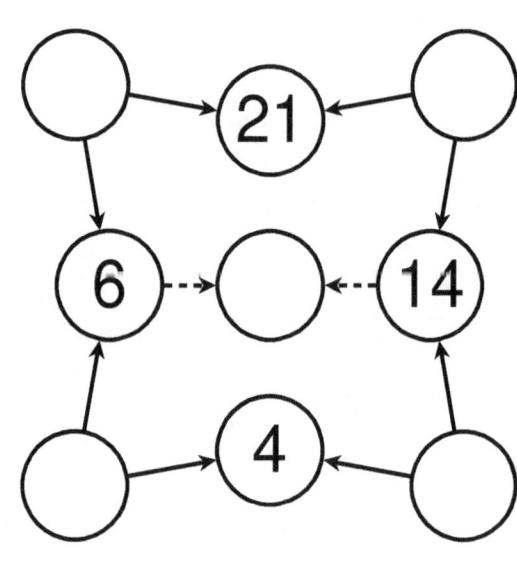

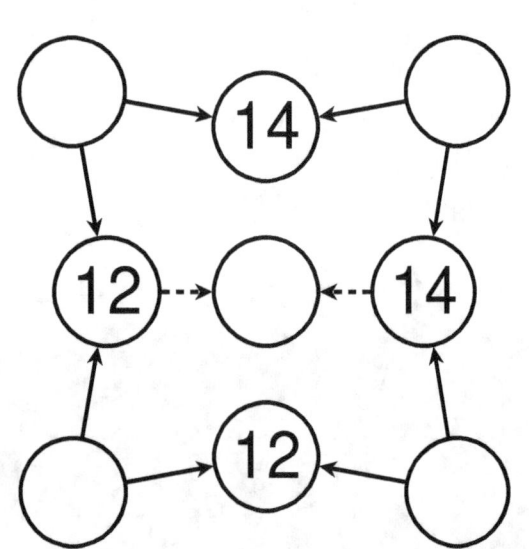

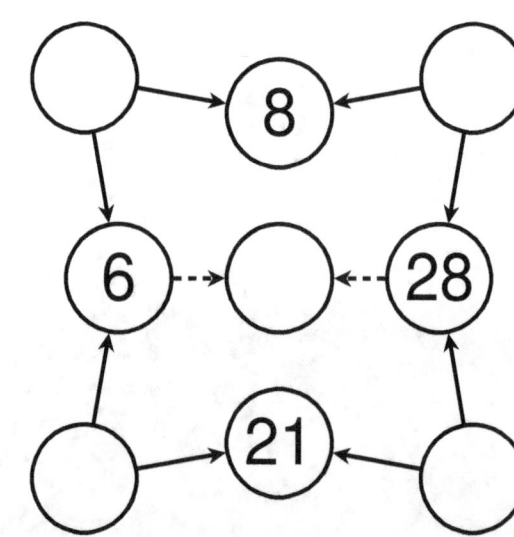

Name _____ Date _____

Find the missing numbers. Solid lines mean multiply.
Dotted lines mean add.

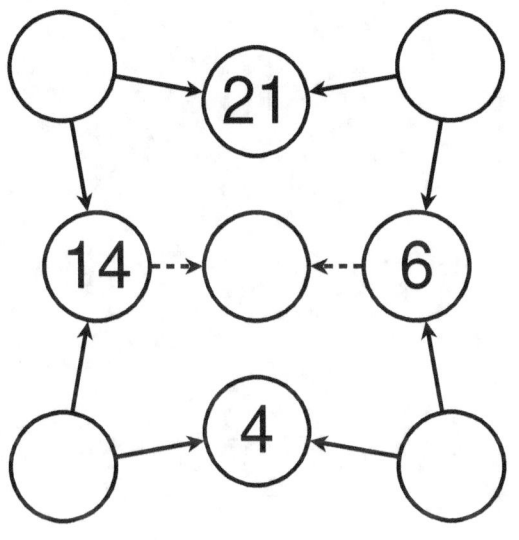

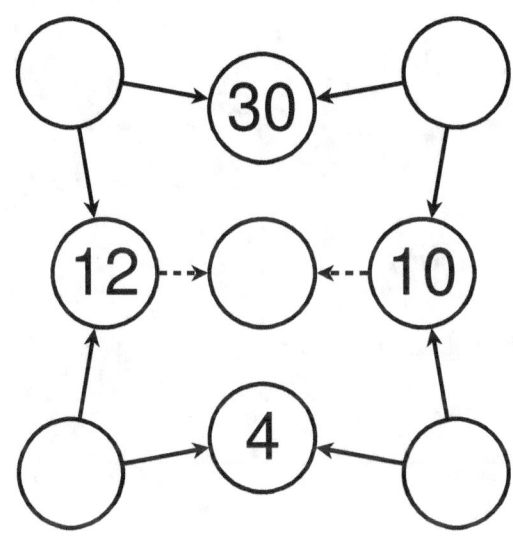

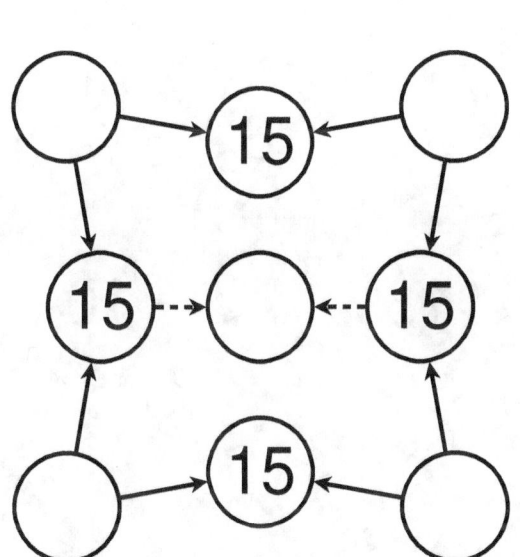

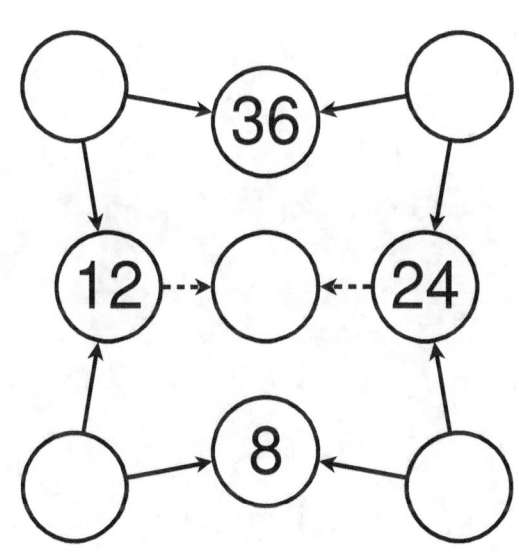

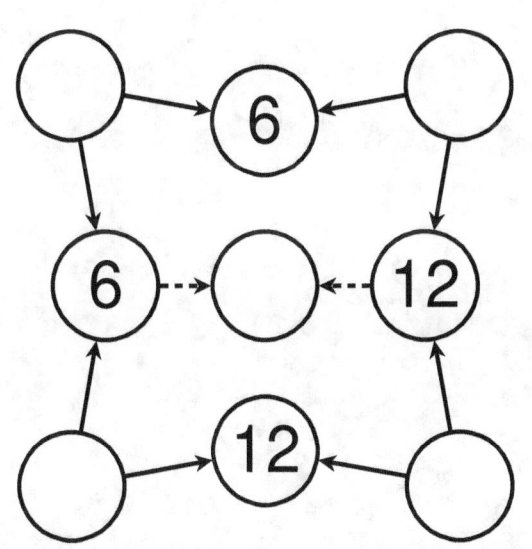

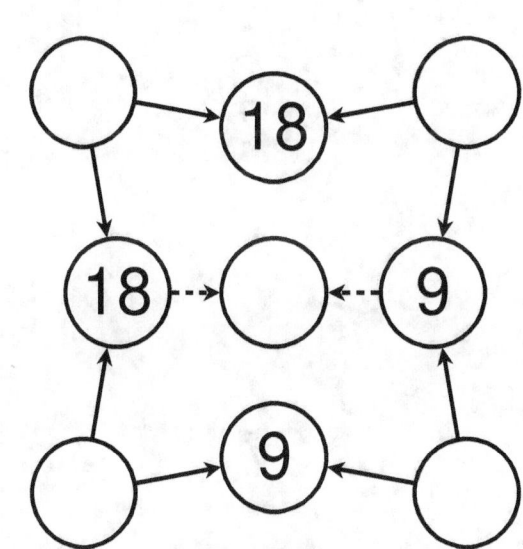

Page 19

Name _____ Date _____

Find the missing numbers. Solid lines mean multiply.
Dotted lines mean add.

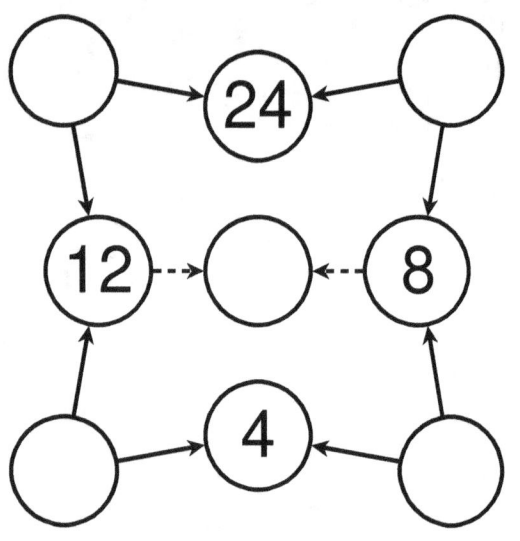

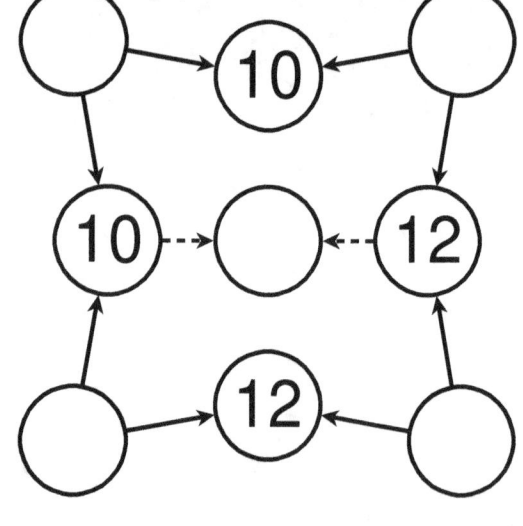

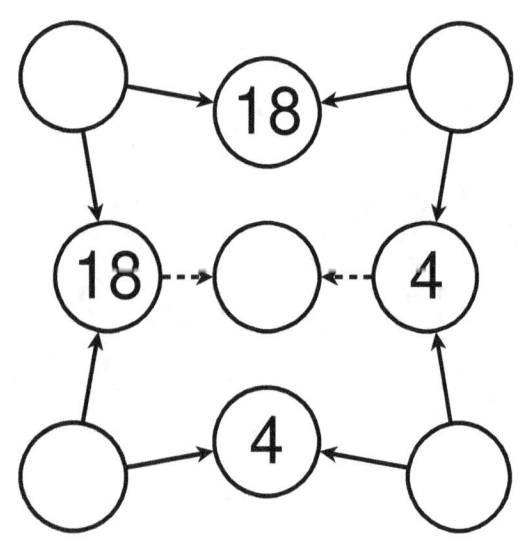

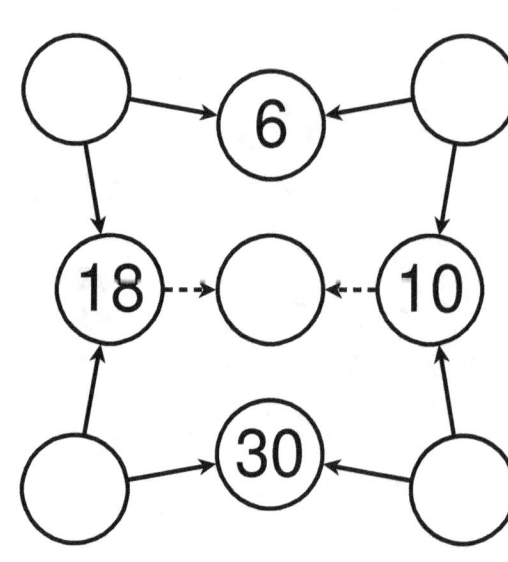

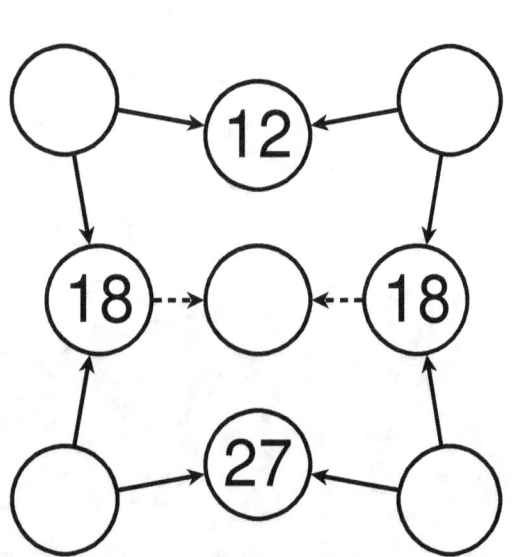

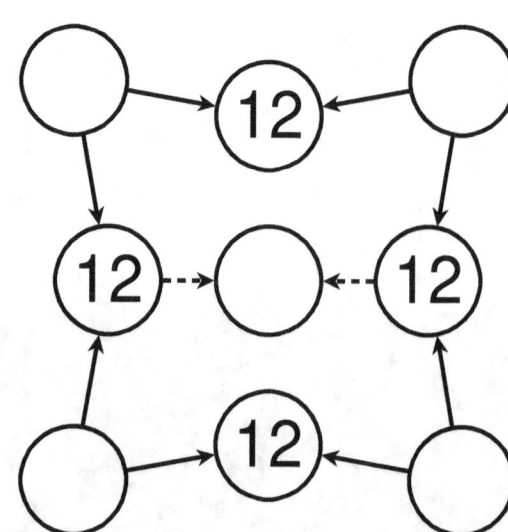

Page 20

Name _____ Date _____

Find the missing numbers. Solid lines mean multiply.
Dotted lines mean add.

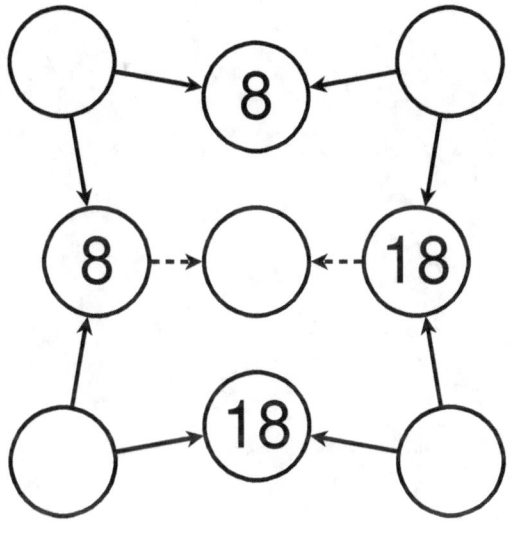

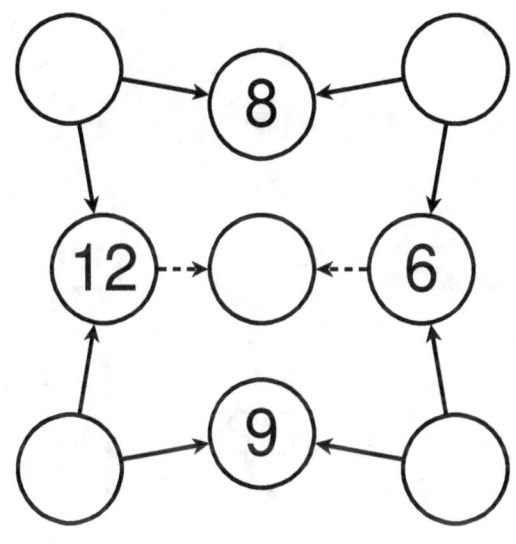

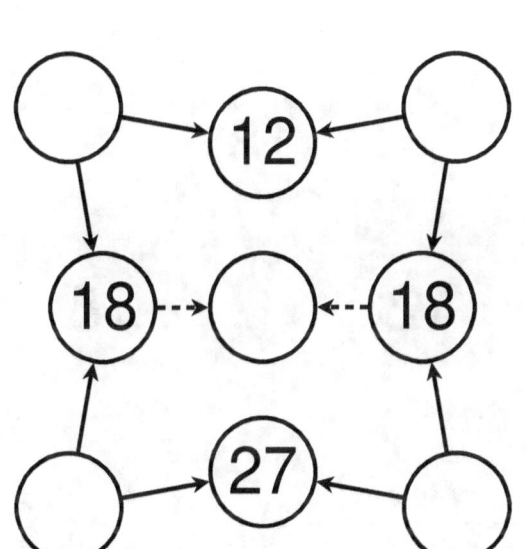

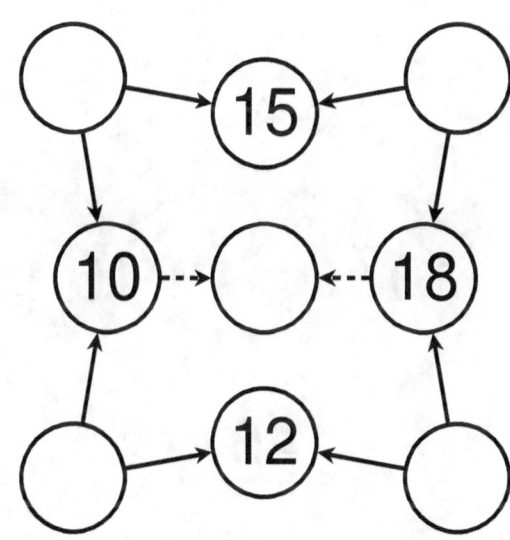

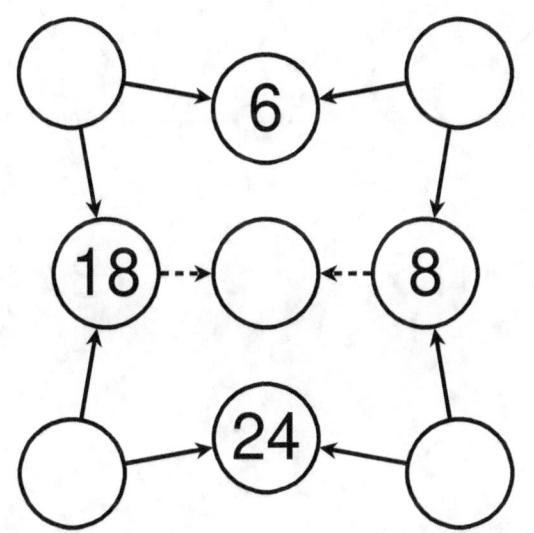

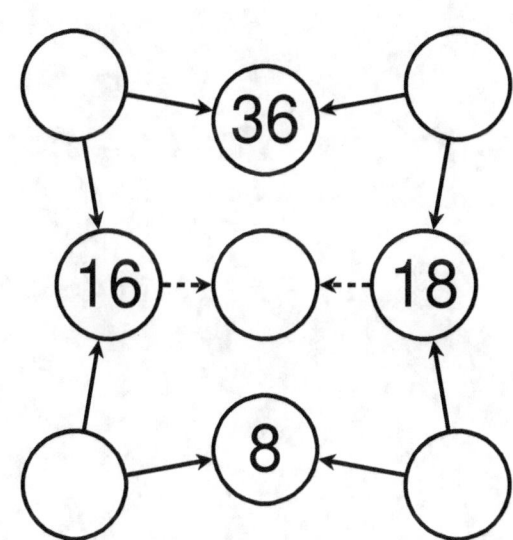

Page 21

Find the missing numbers. Solid lines mean multiply.
Dotted lines mean add.

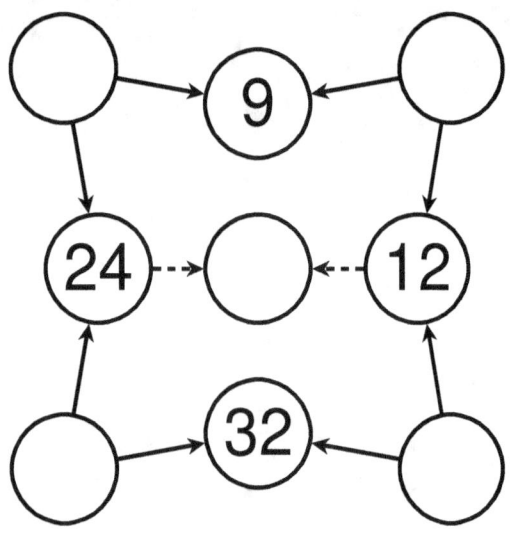

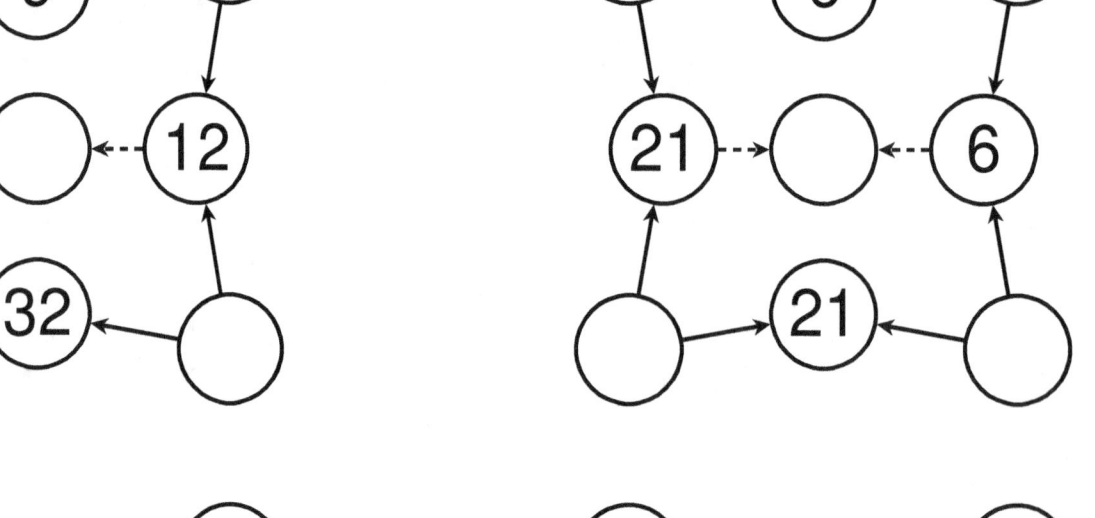

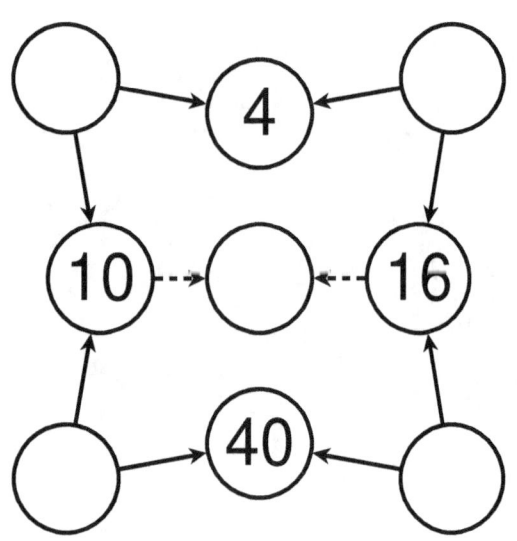

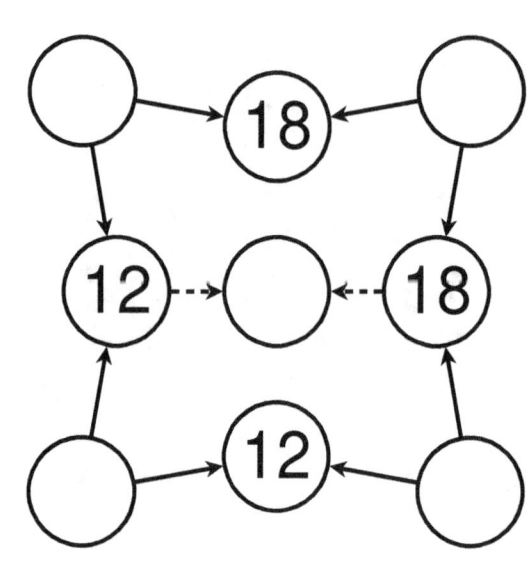

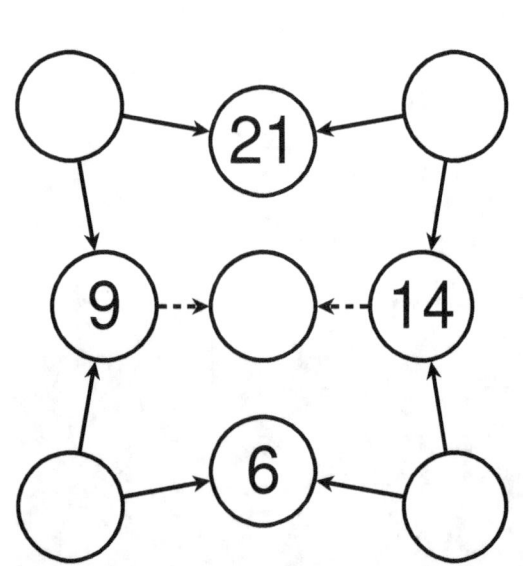

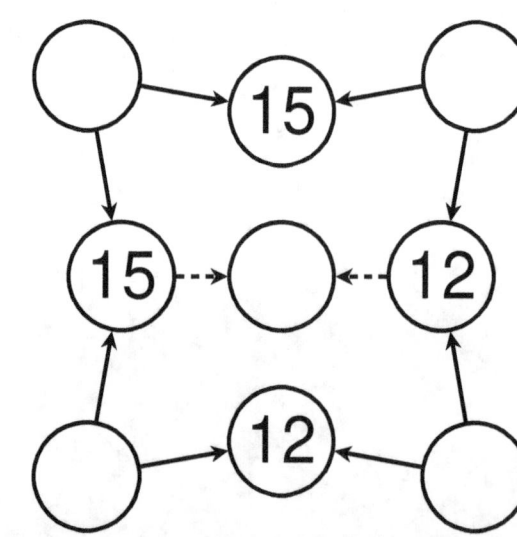

Name _____   Date _____

Find the missing numbers. Solid lines mean multiply.
Dotted lines mean add.

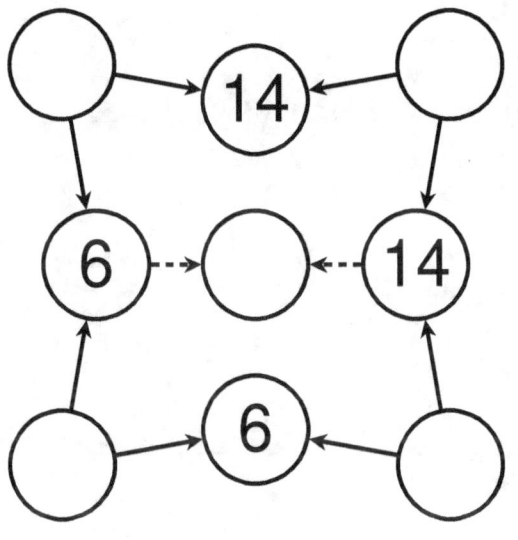

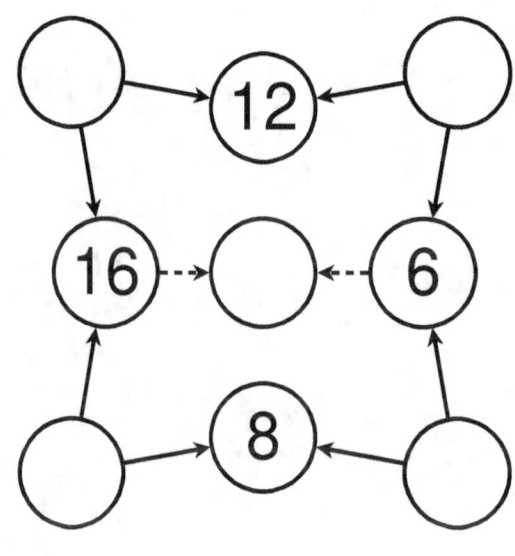

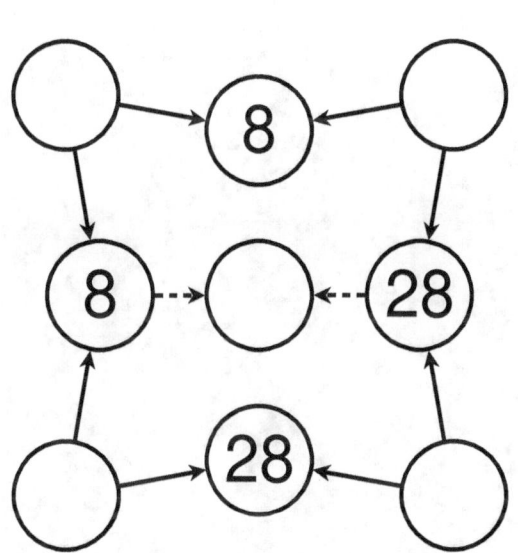

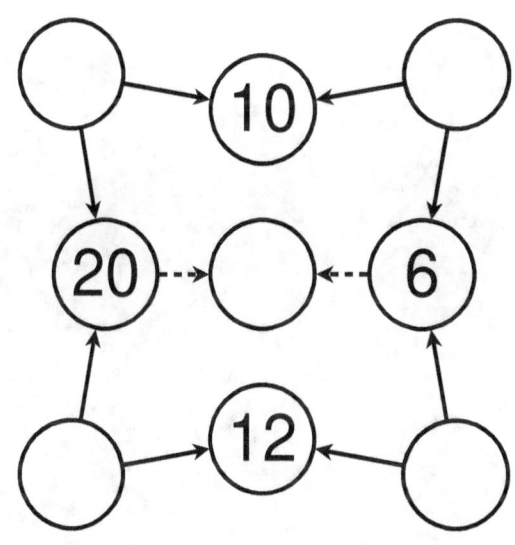

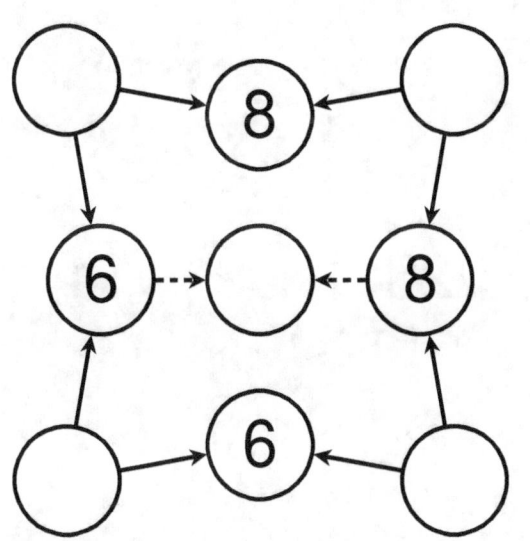

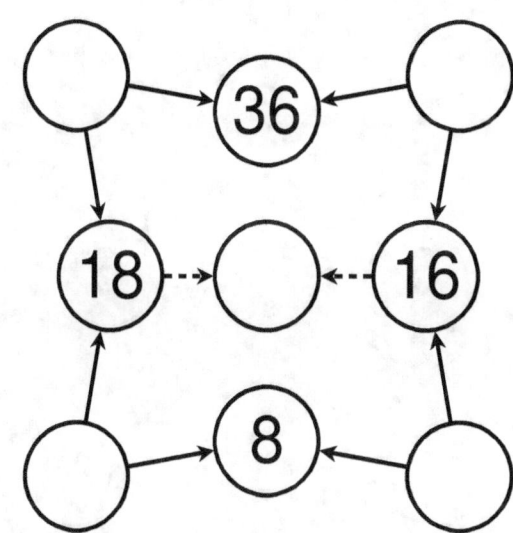

Name _____ Date _____

Find the missing numbers. Solid lines mean multiply.
Dotted lines mean add.

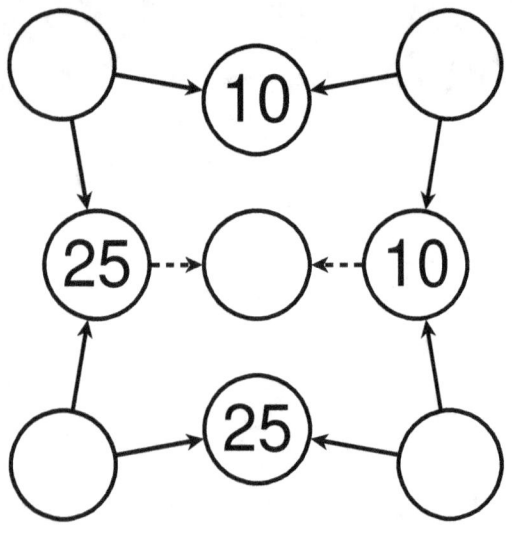
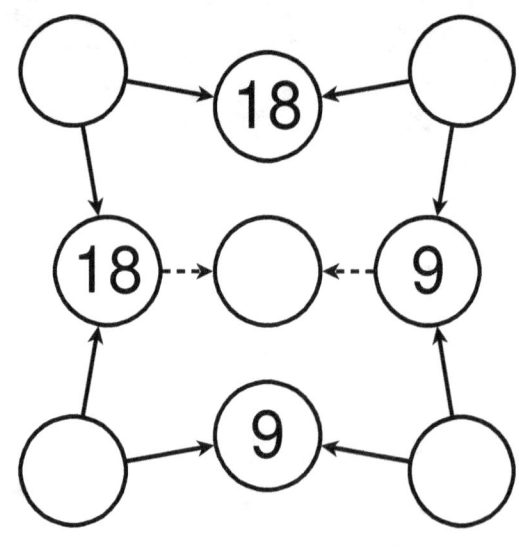

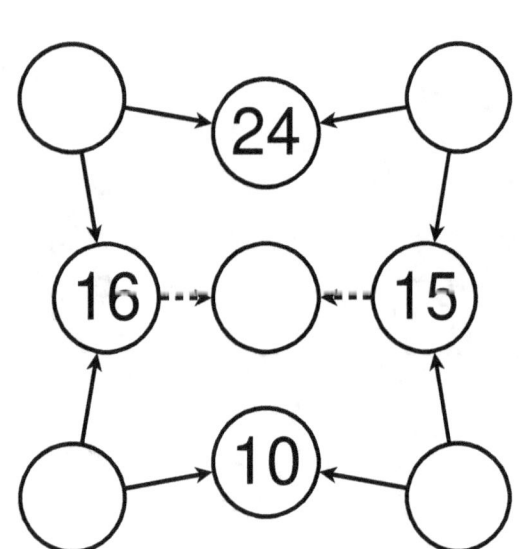
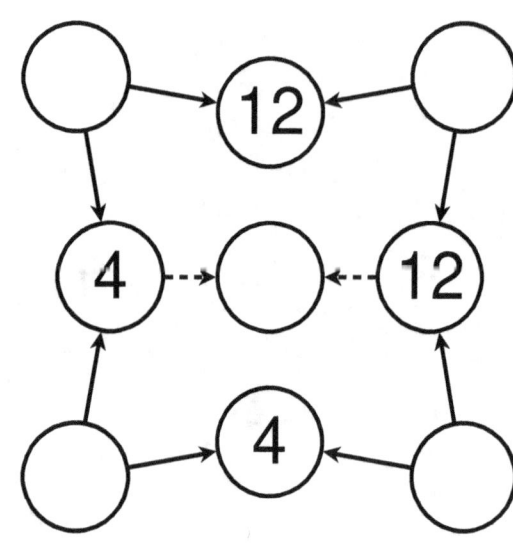

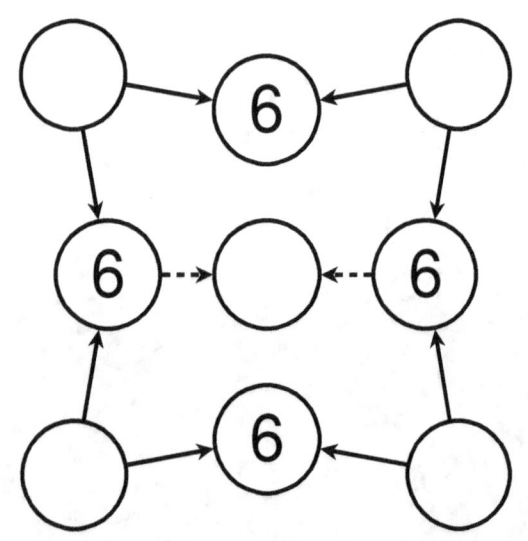
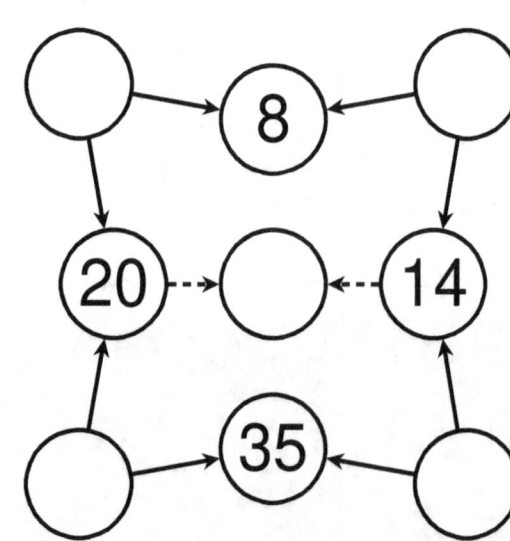

Name _____ Date _____

Find the missing numbers. Solid lines mean multiply.
Dotted lines mean add.

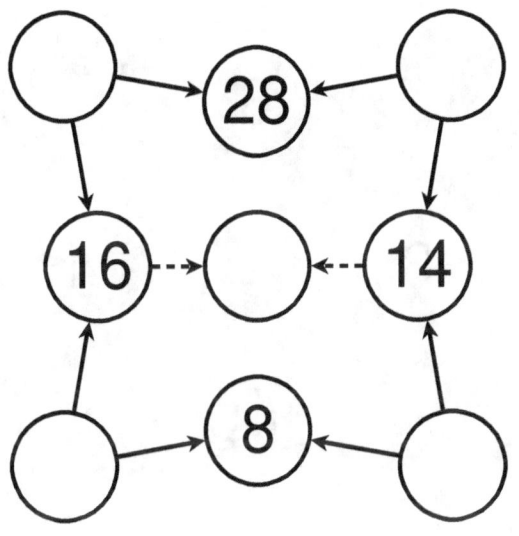

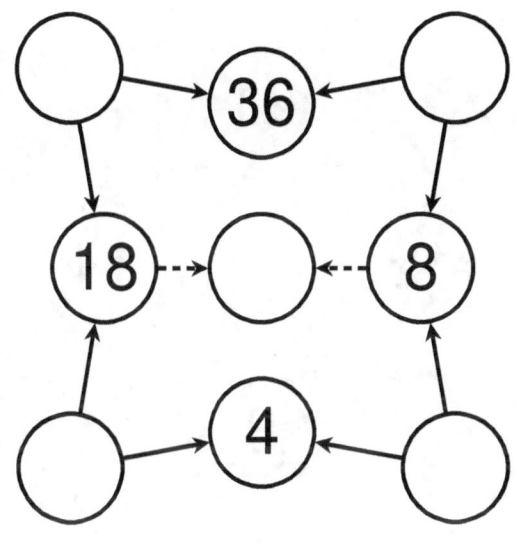

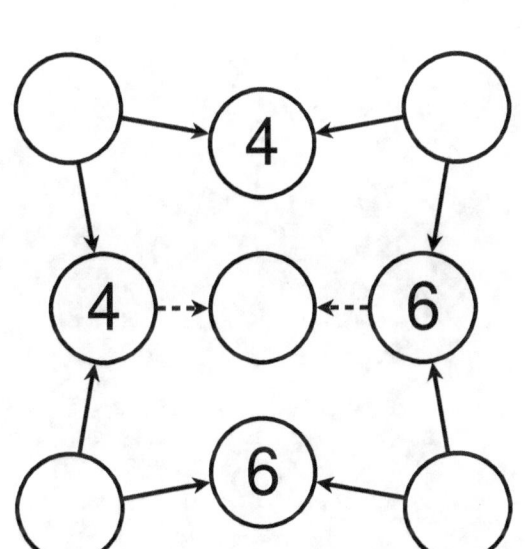

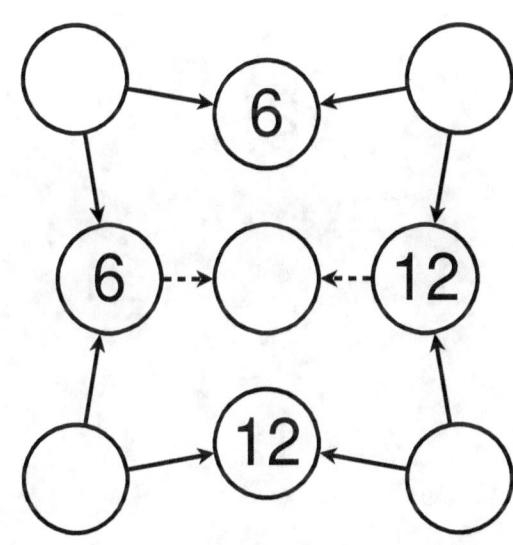

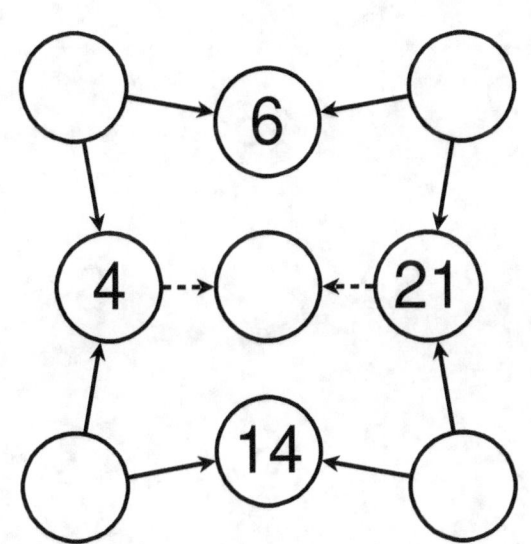

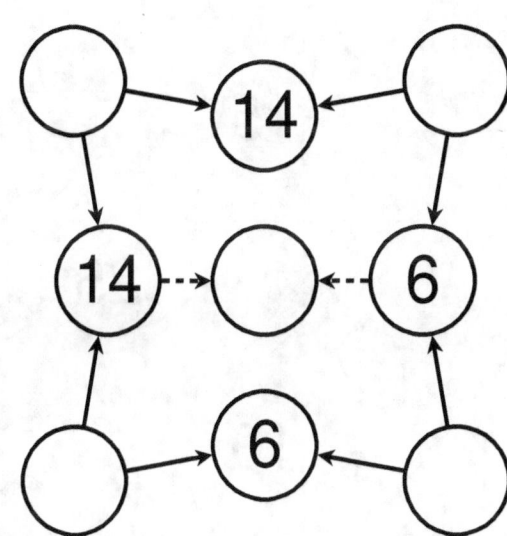

Name _____ Date _____

Find the missing numbers. Solid lines mean multiply.
Dotted lines mean add.

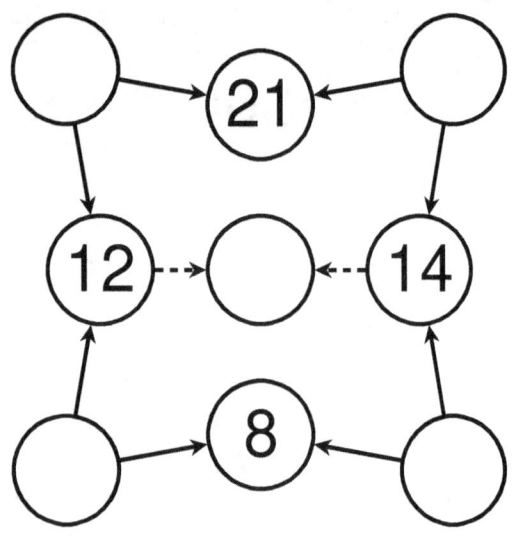

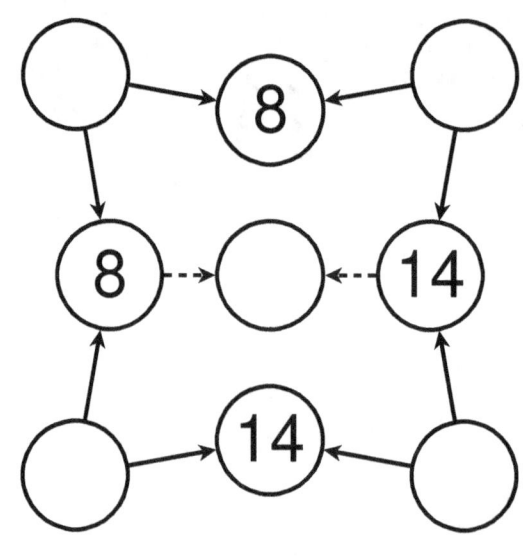

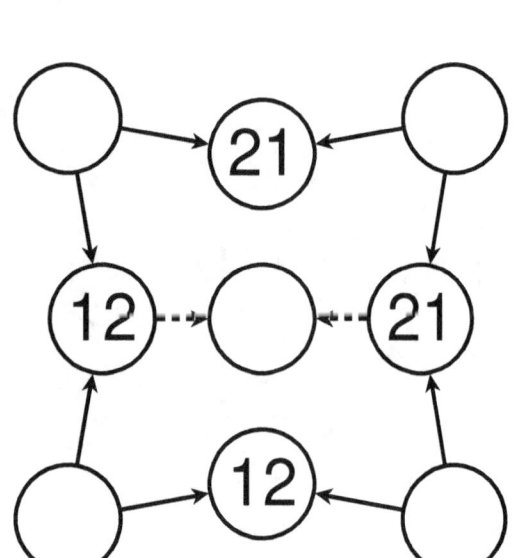

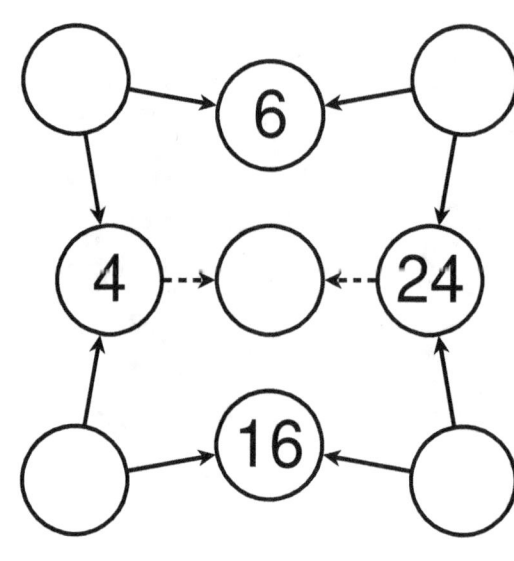

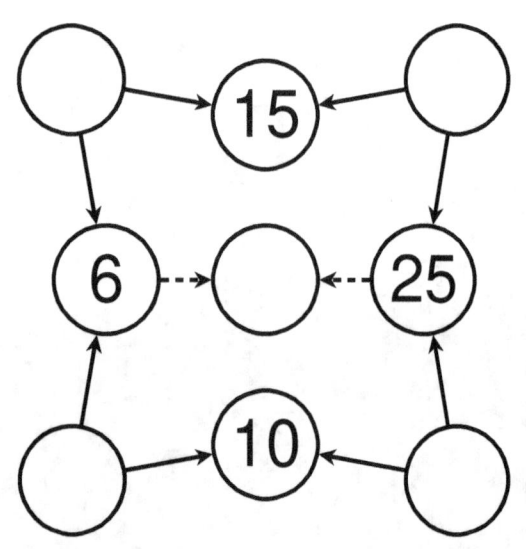

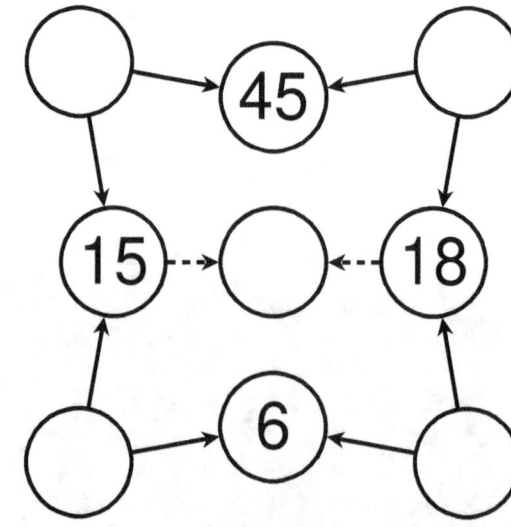

Name _____ Date _____

Find the missing numbers. Solid lines mean multiply.
Dotted lines mean add.

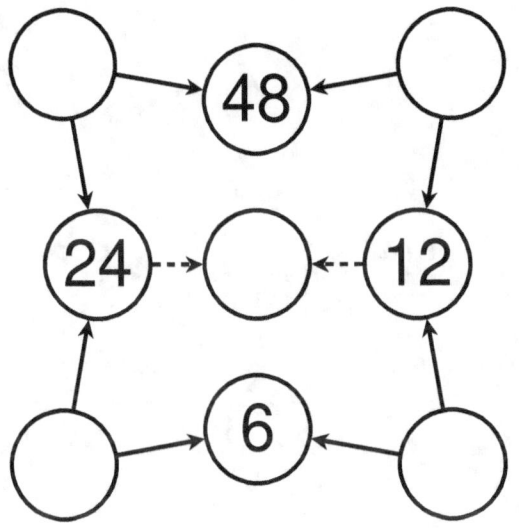

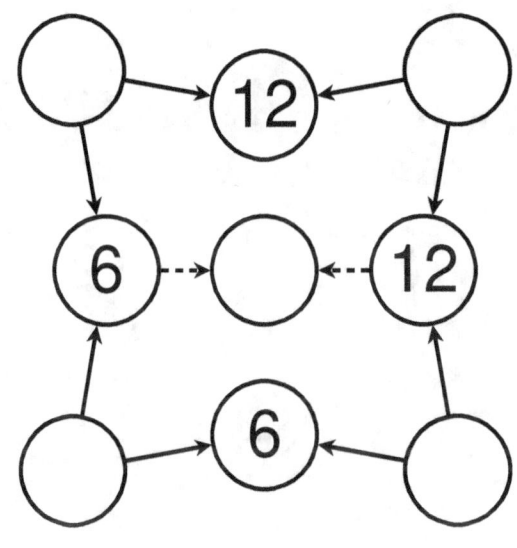

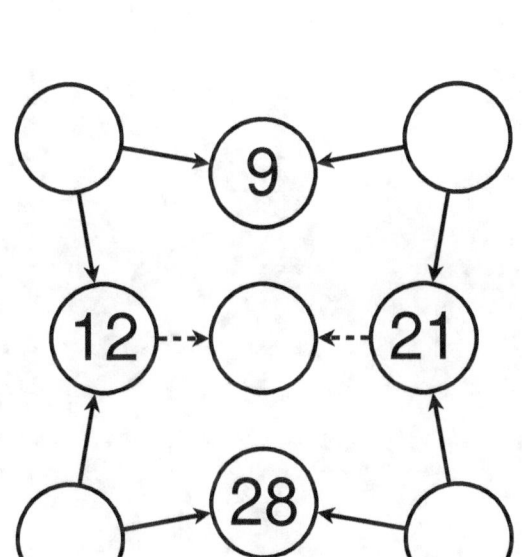

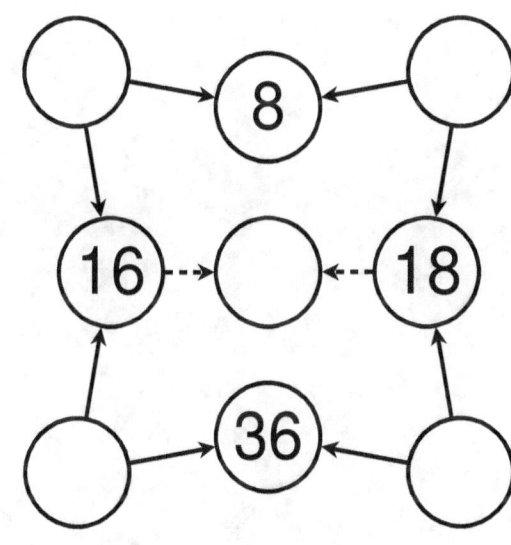

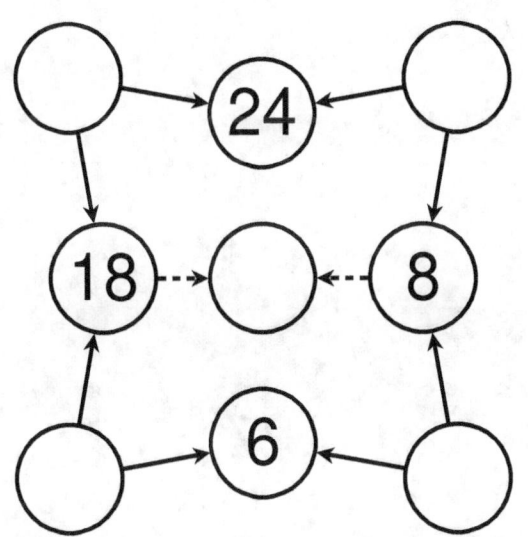

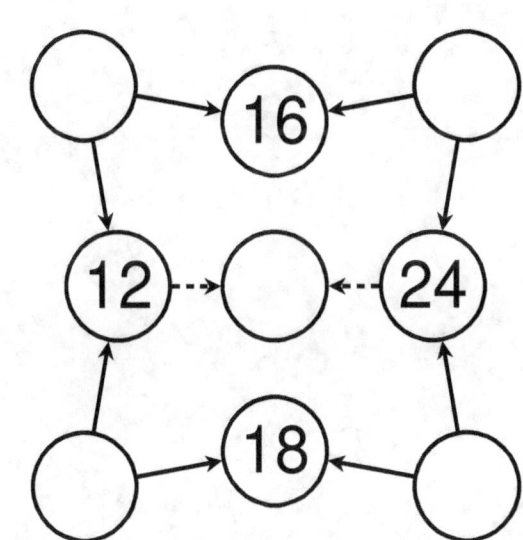

Name _____ Date _____

Find the missing numbers. Solid lines mean multiply.
Dotted lines mean add.

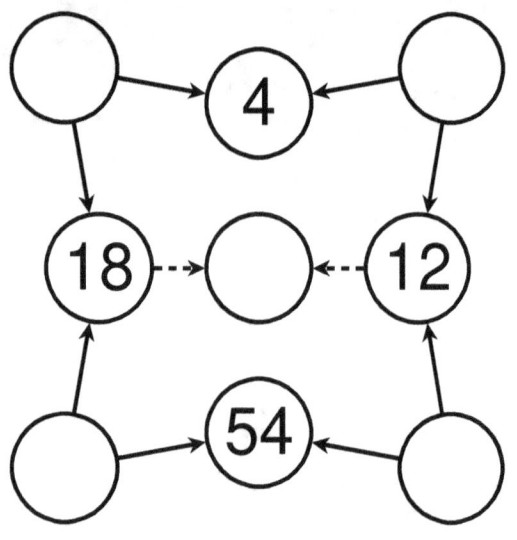

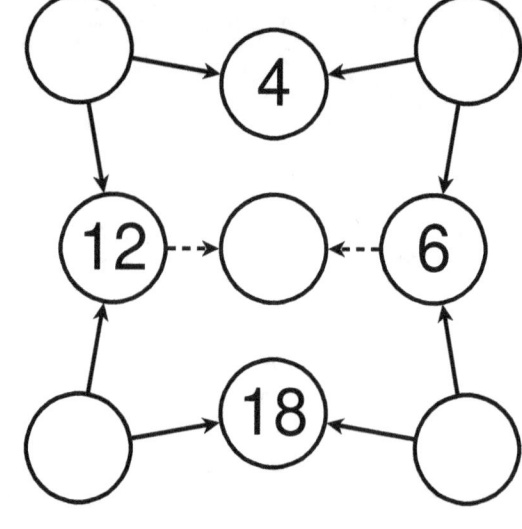

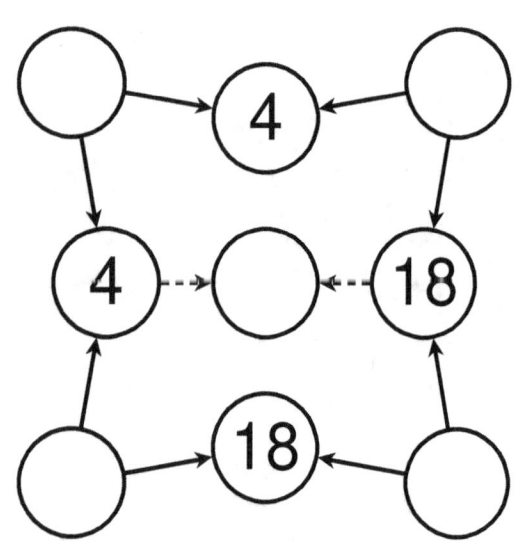

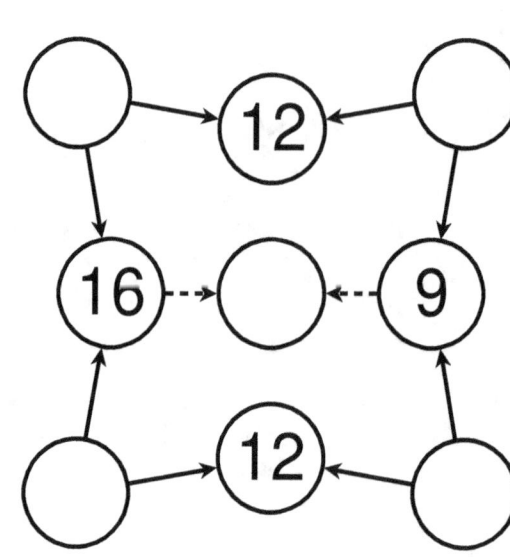

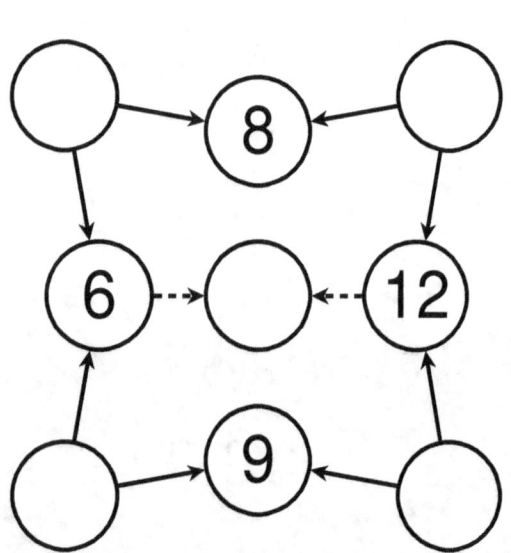

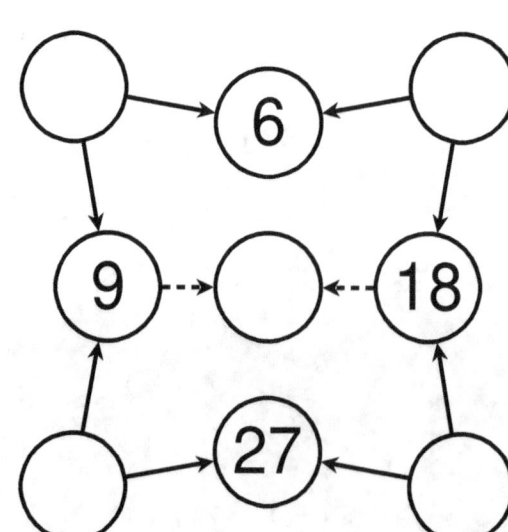

Name _____ Date _____

Find the missing numbers. Solid lines mean multiply.
Dotted lines mean add.

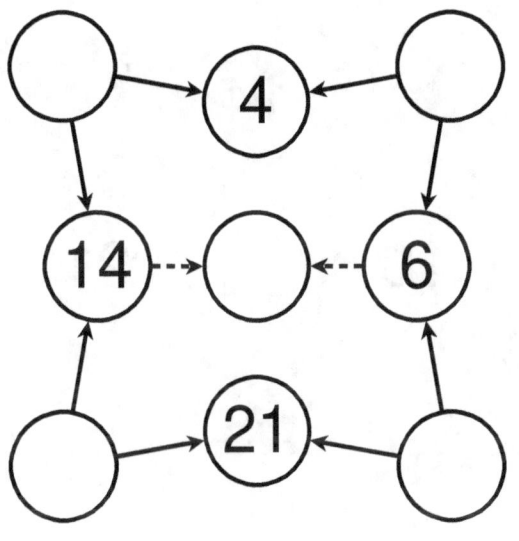

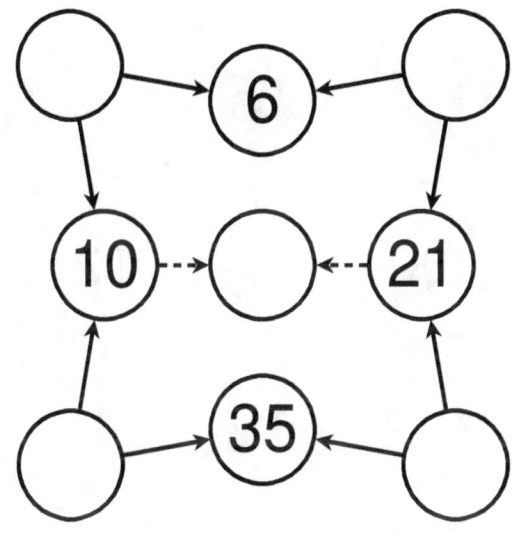

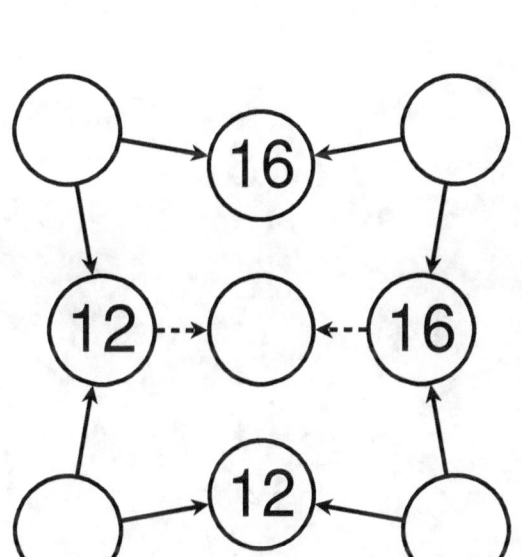

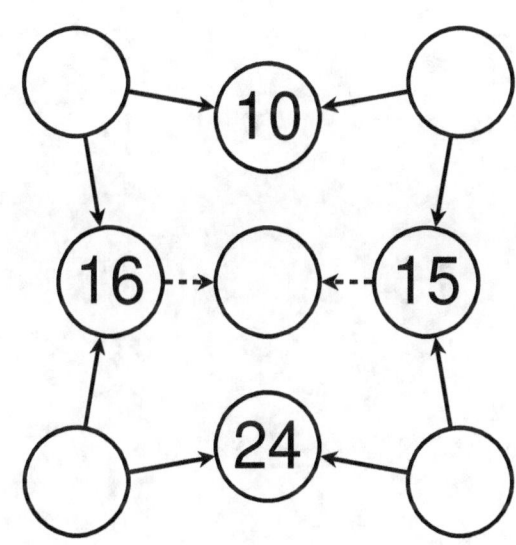

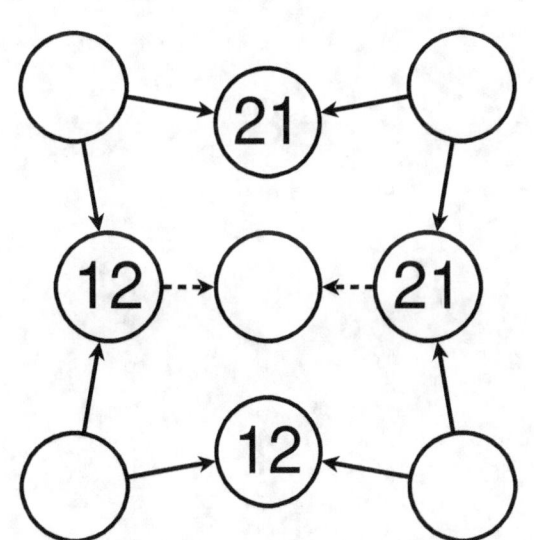

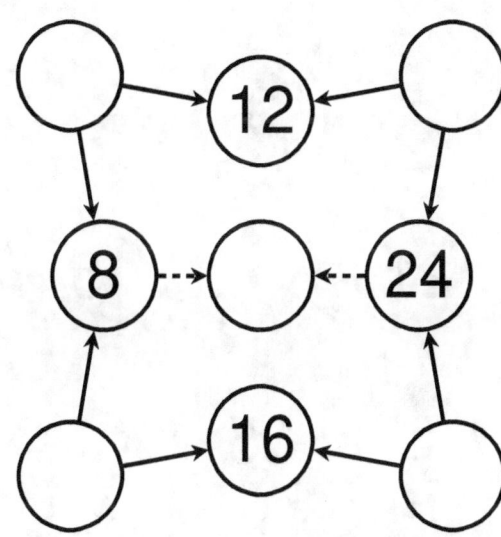

Name _____ Date _____

Find the missing numbers. Solid lines mean multiply.
Dotted lines mean add.

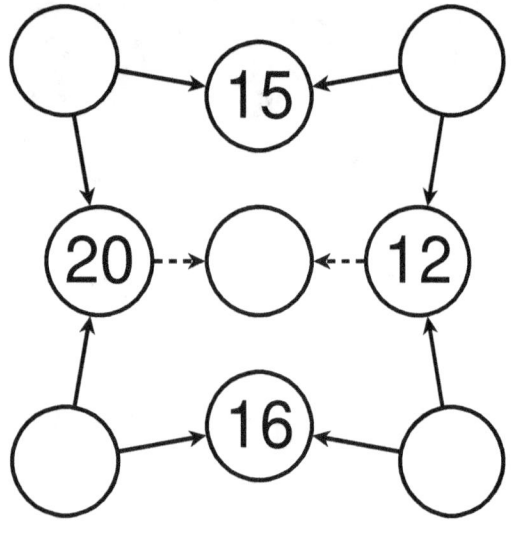
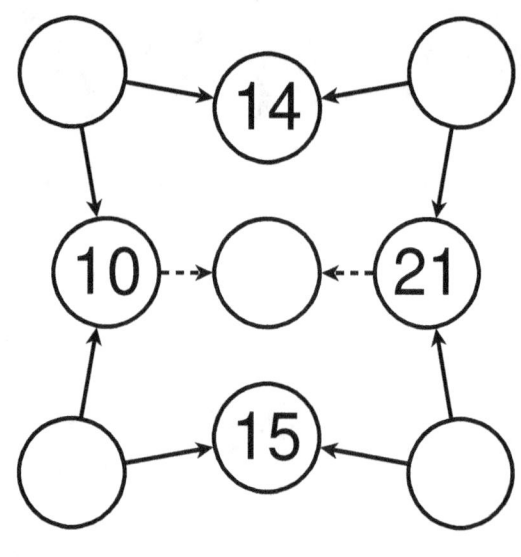
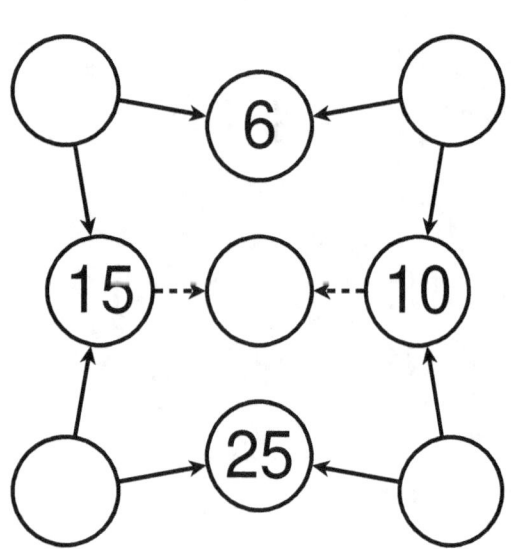
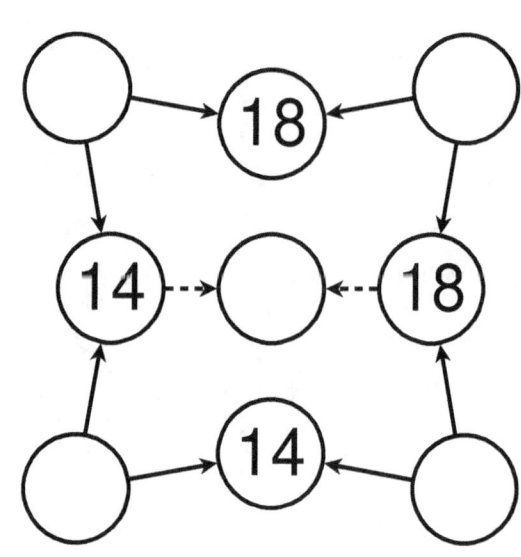
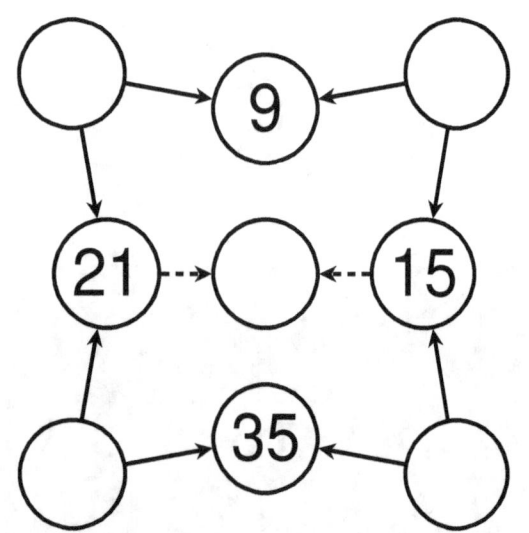
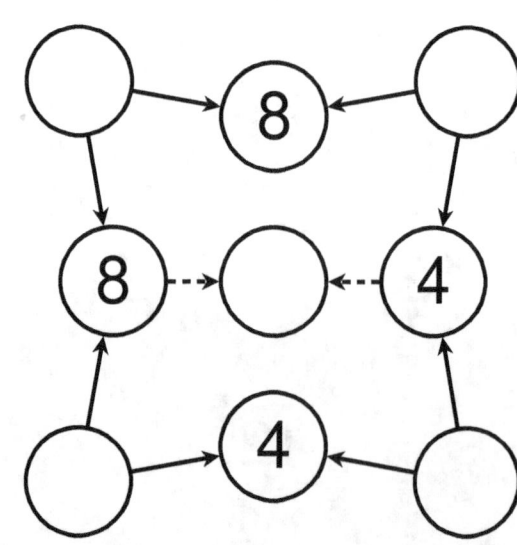

Page 30

Name _____ Date _____

Find the missing numbers. Solid lines mean multiply.
Dotted lines mean add.

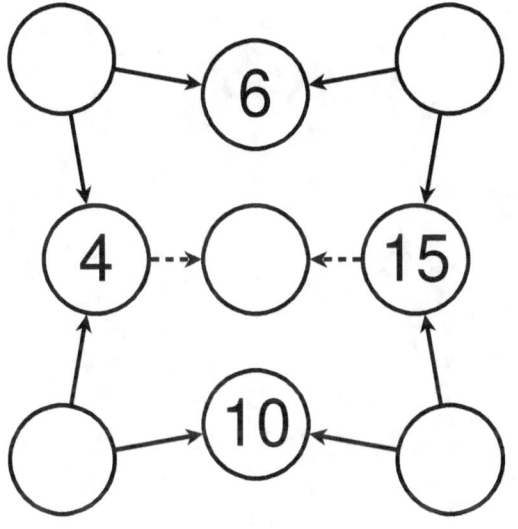

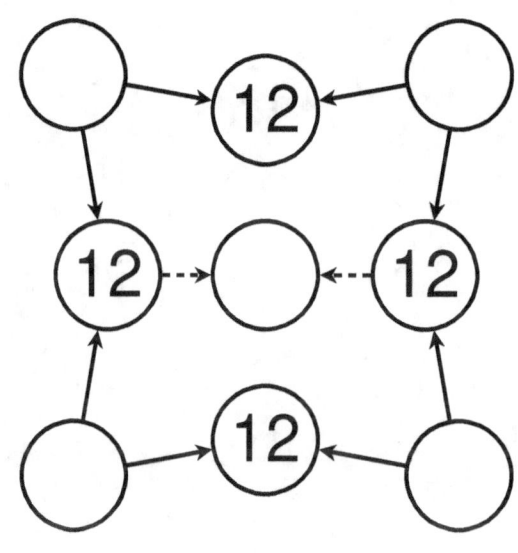

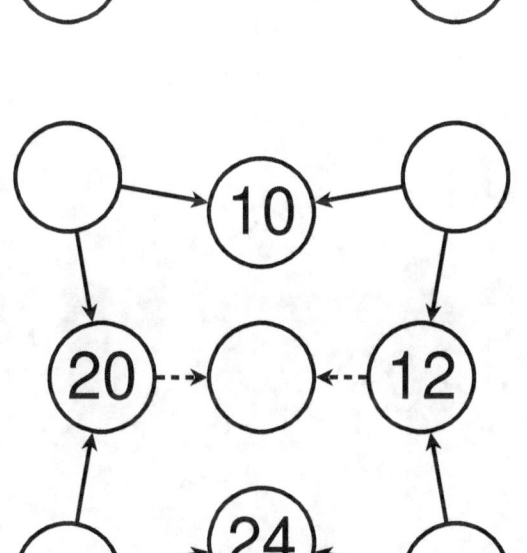

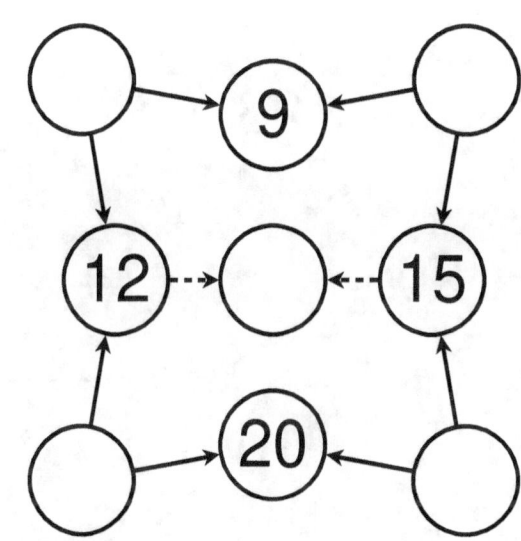

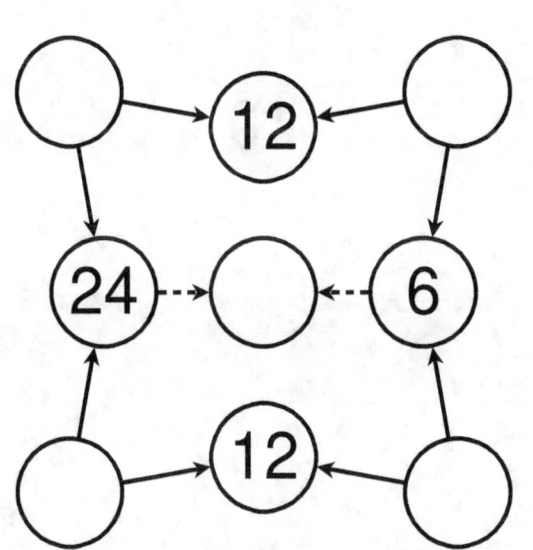

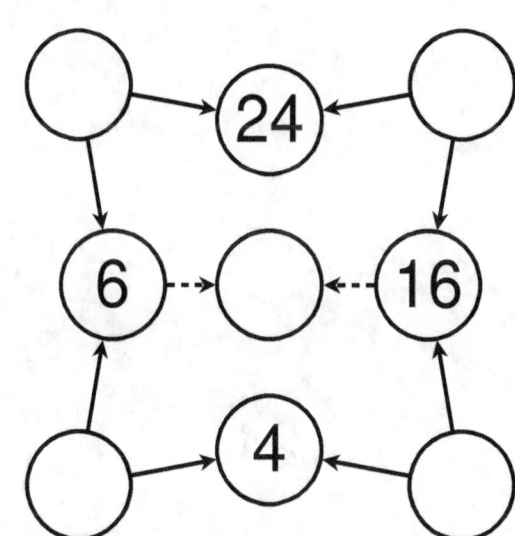

Name _____ Date _____

Find the missing numbers. Solid lines mean multiply.
Dotted lines mean add.

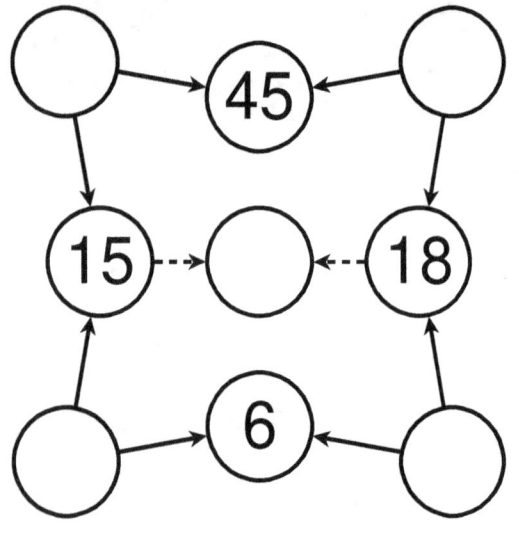

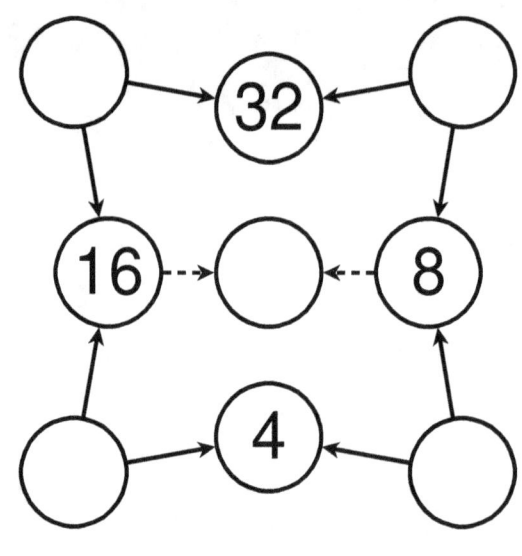

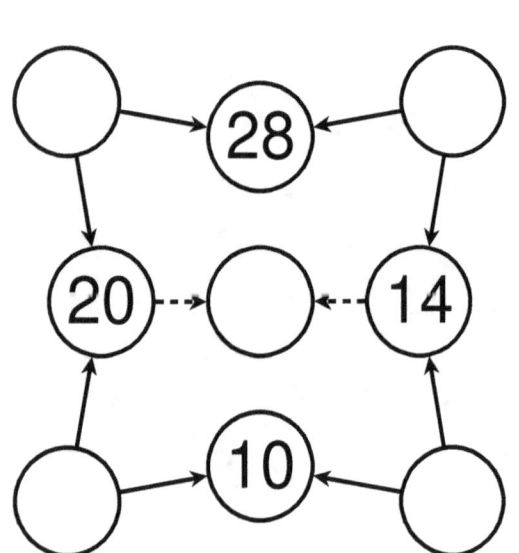

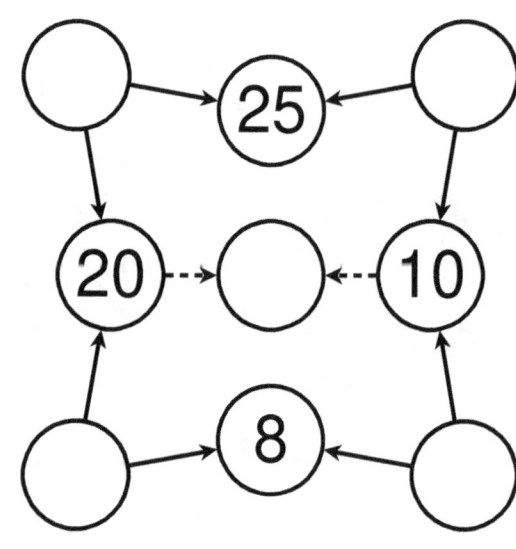

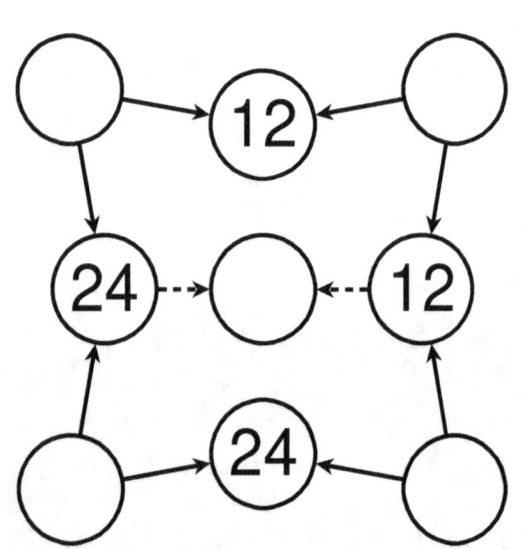

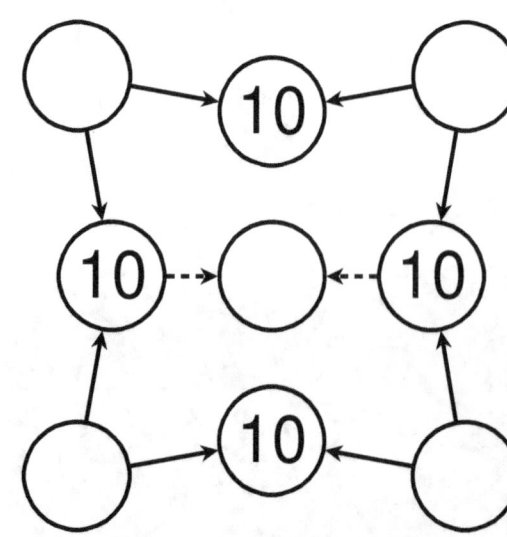

Name _____   Date _____

Find the missing numbers. Solid lines mean multiply.
Dotted lines mean add.

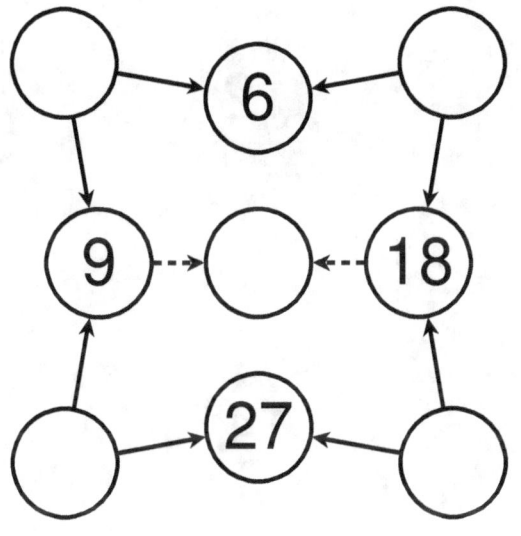
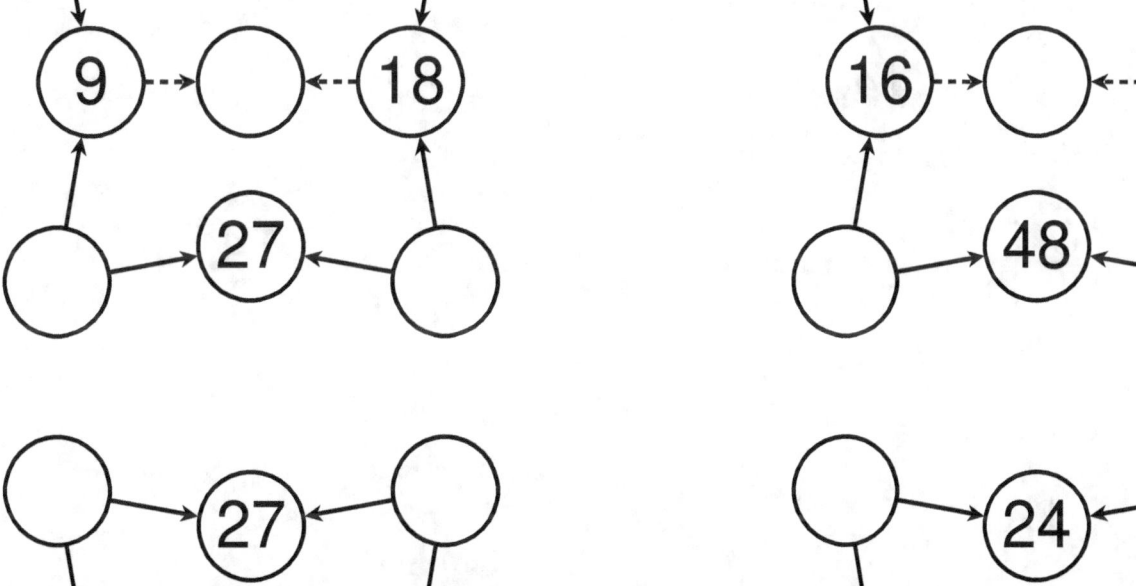
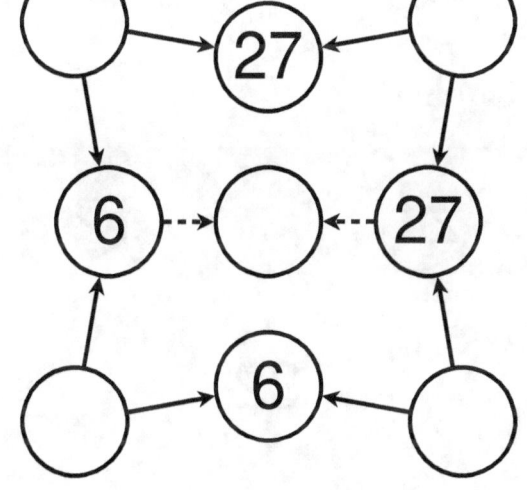
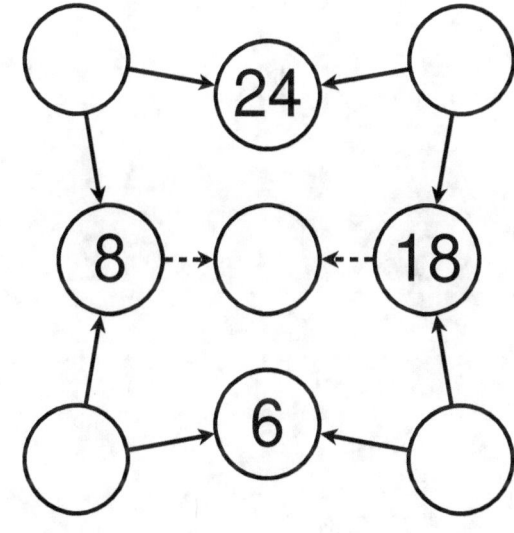
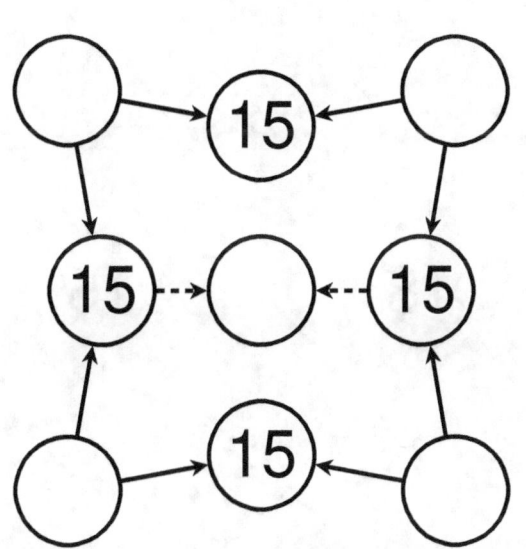
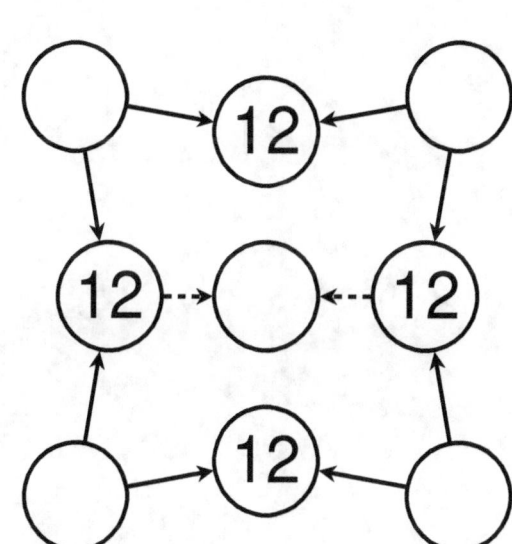

Page 33

Name _____ Date _____

Find the missing numbers. Solid lines mean multiply.
Dotted lines mean add.

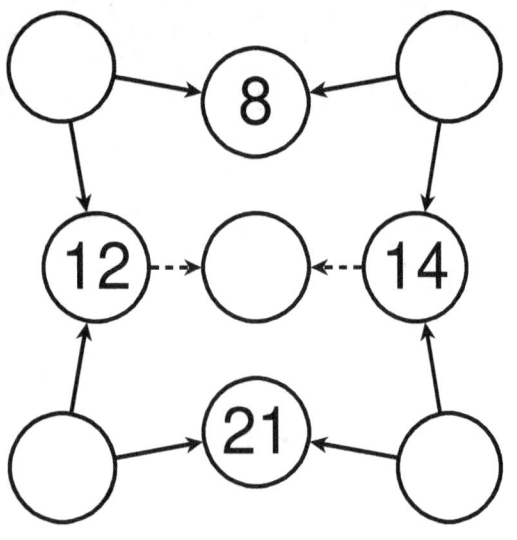

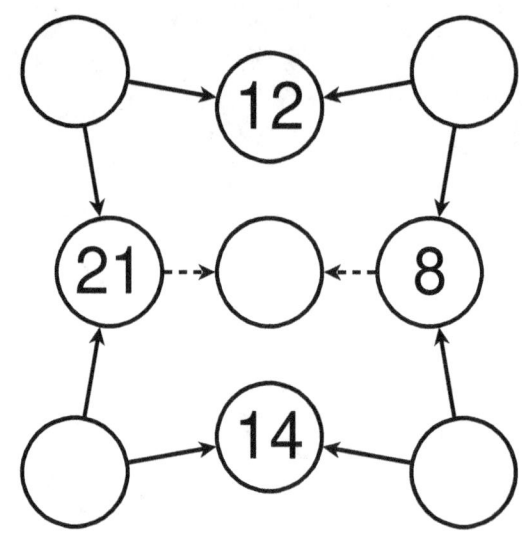

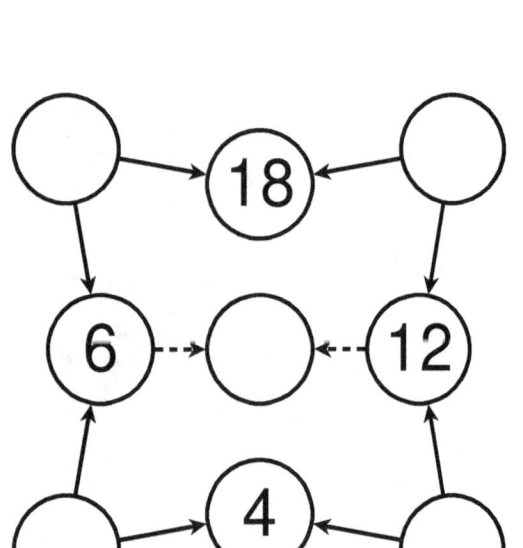

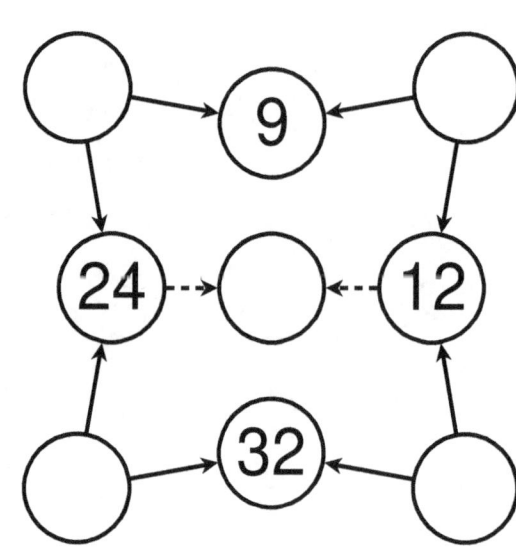

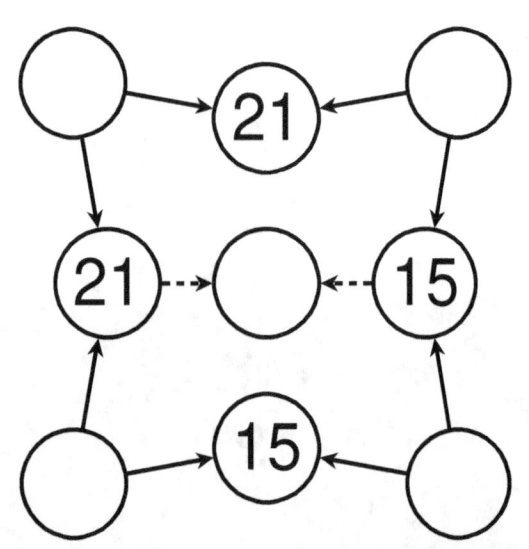

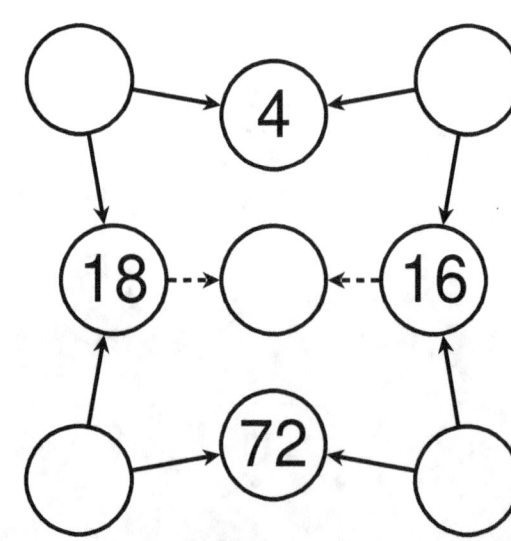

Page 34

Name _____ Date _____

Find the missing numbers. Solid lines mean multiply.
Dotted lines mean add.

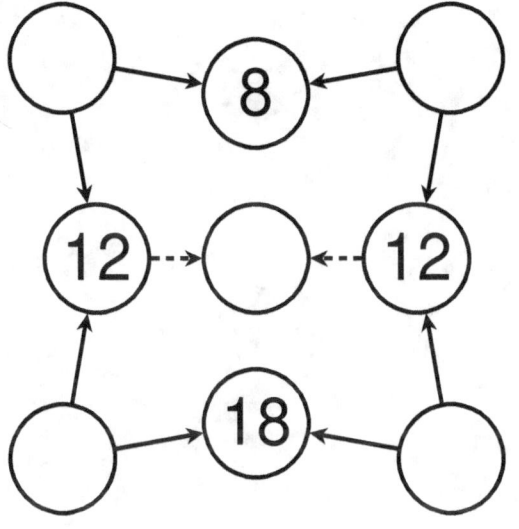

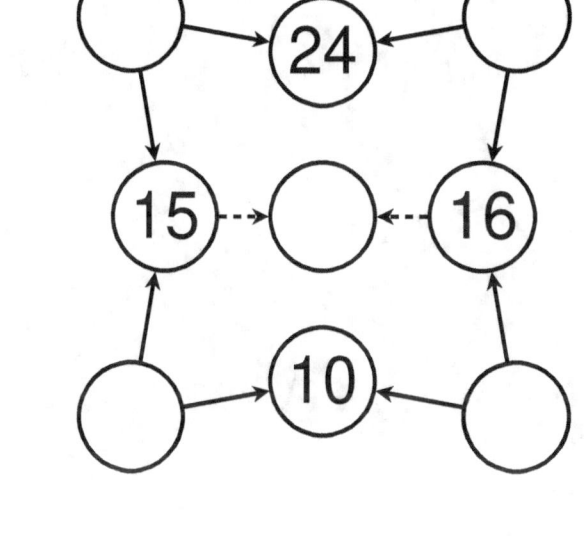

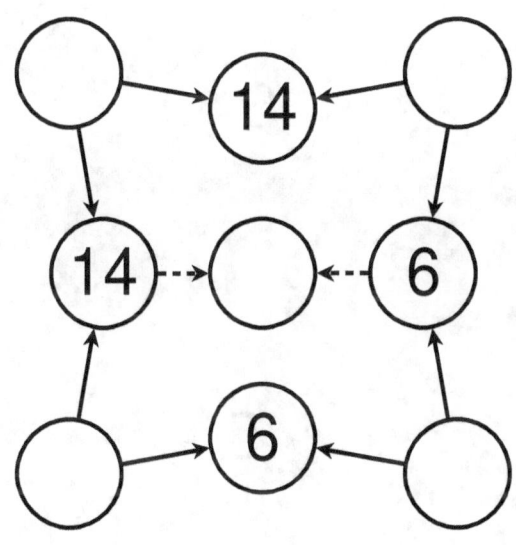

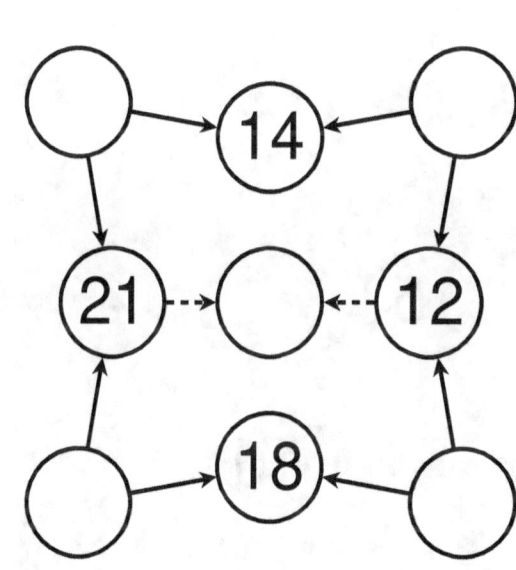

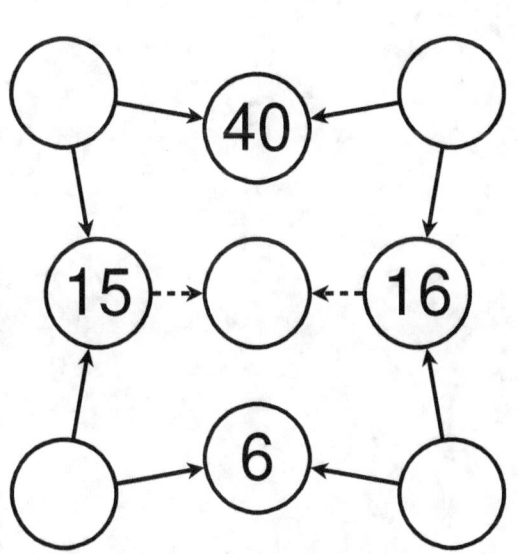

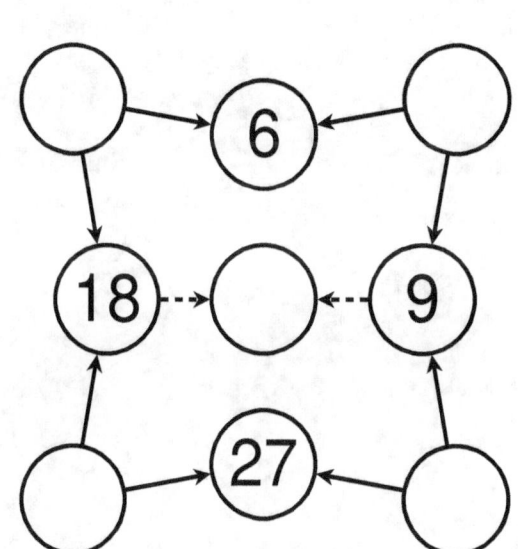

Name _____ Date _____

Find the missing numbers. Solid lines mean multiply.
Dotted lines mean add.

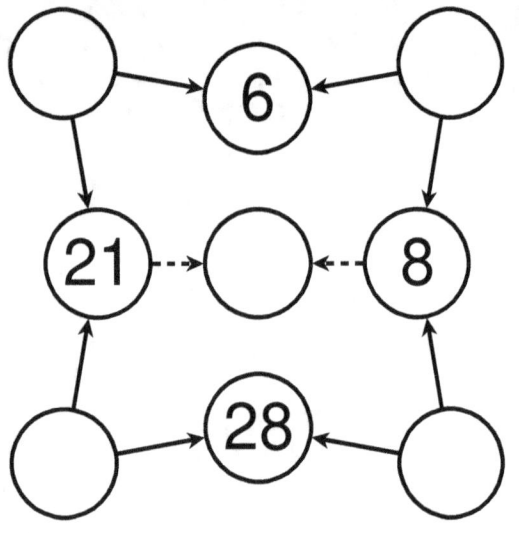

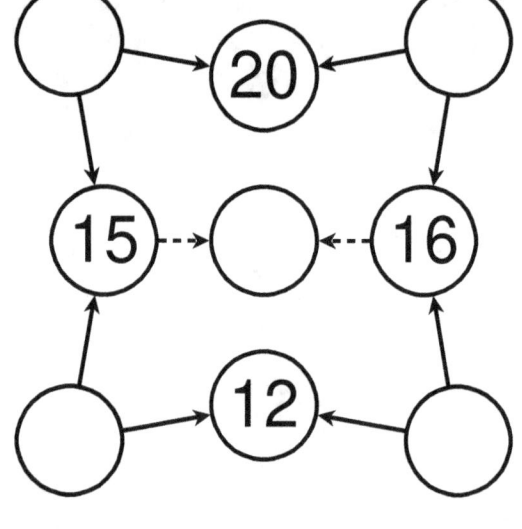

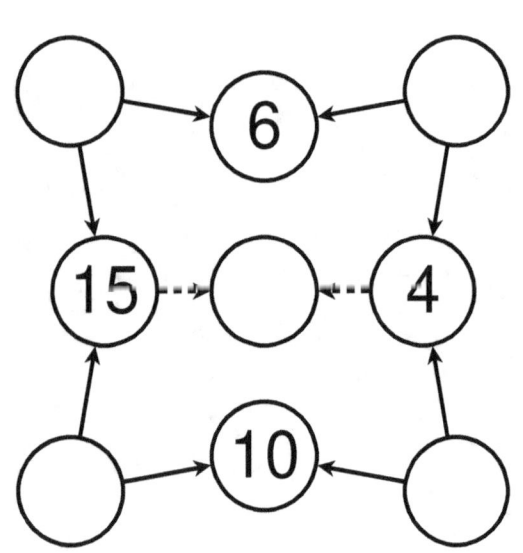

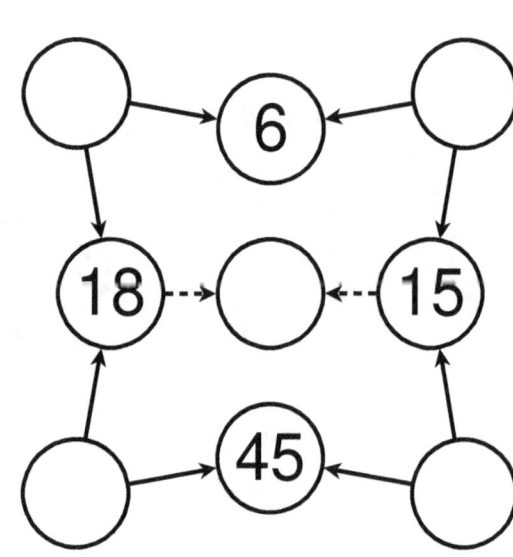

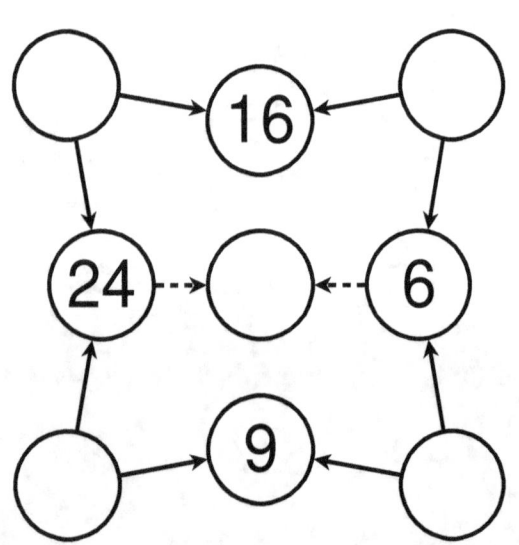

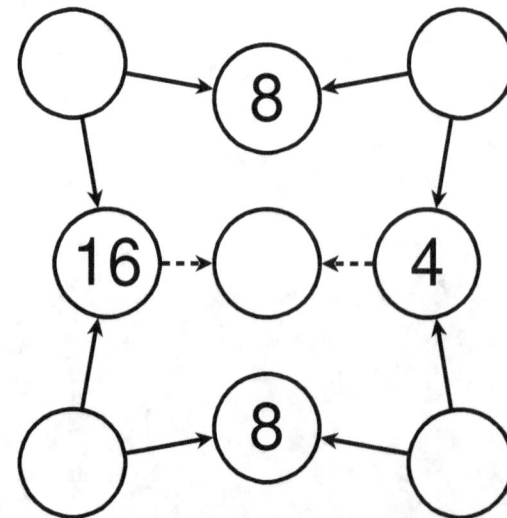

Page 36

Name _____   Date _____

Find the missing numbers. Solid lines mean multiply.
Dotted lines mean add.

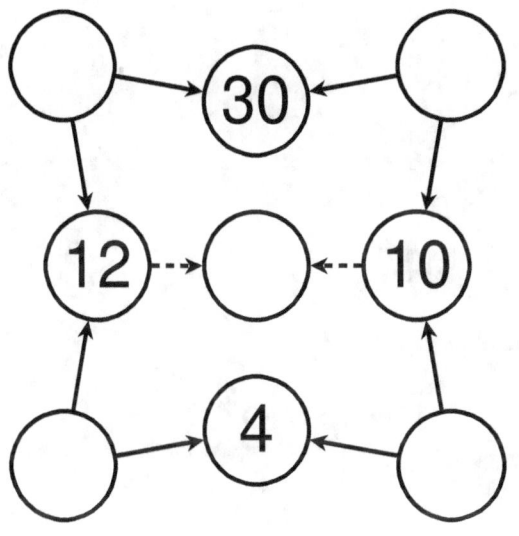

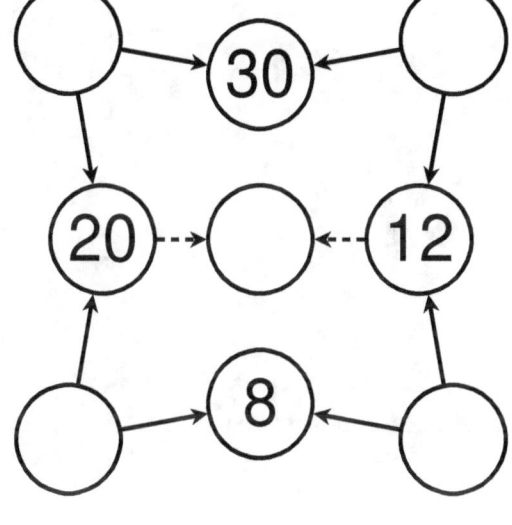

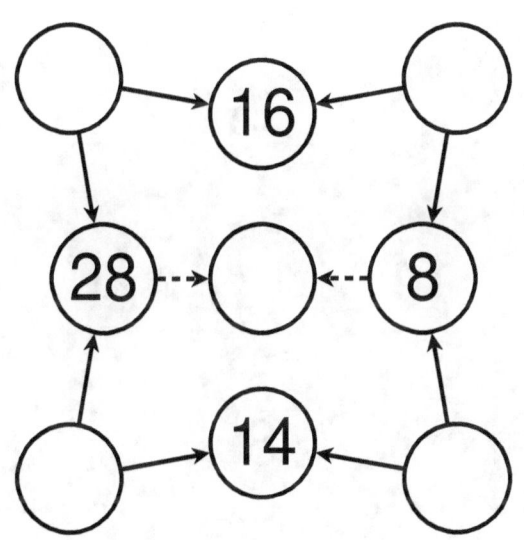

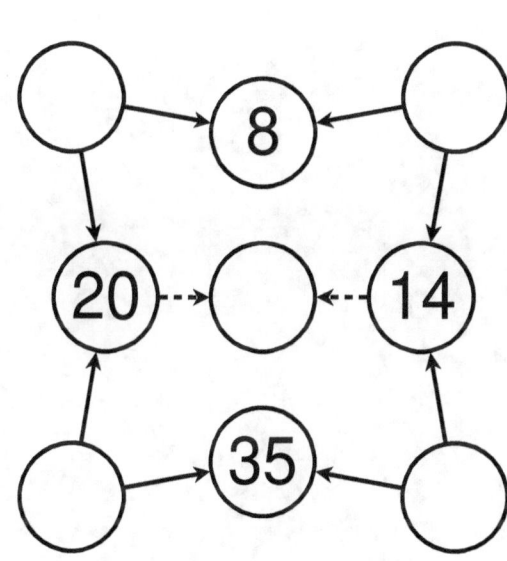

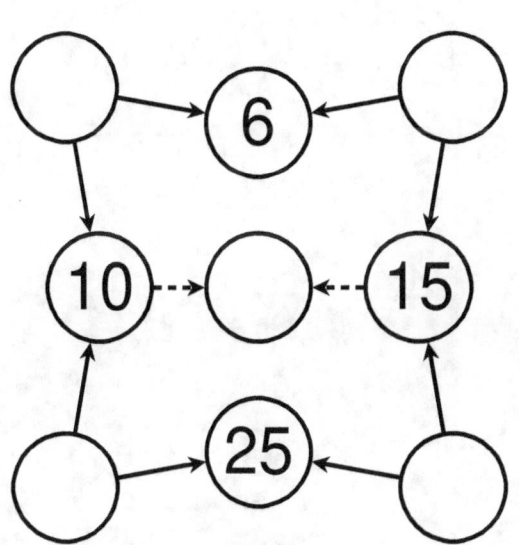

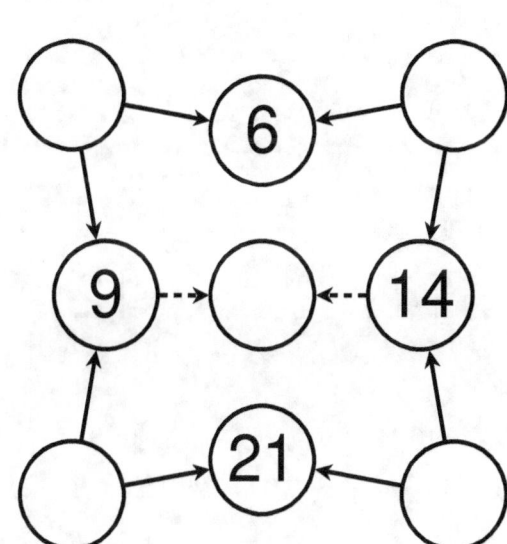

Name _____ Date _____

Find the missing numbers. Solid lines mean multiply.
Dotted lines mean add.

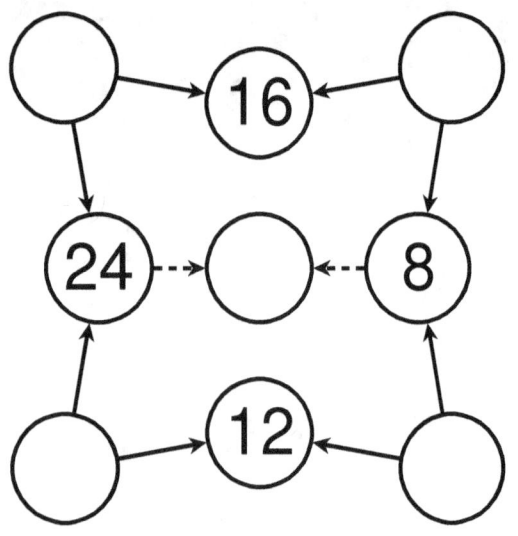

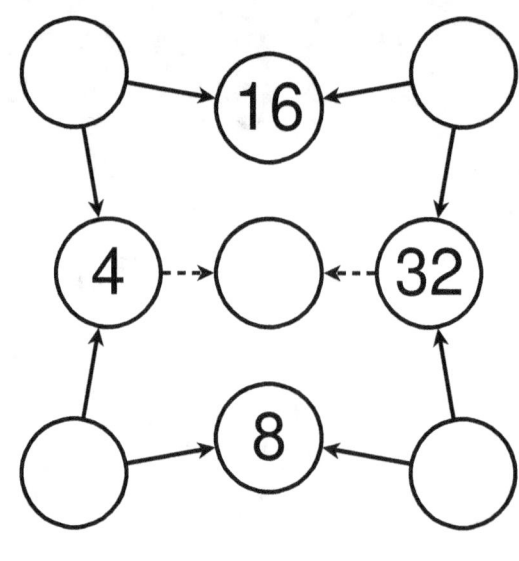

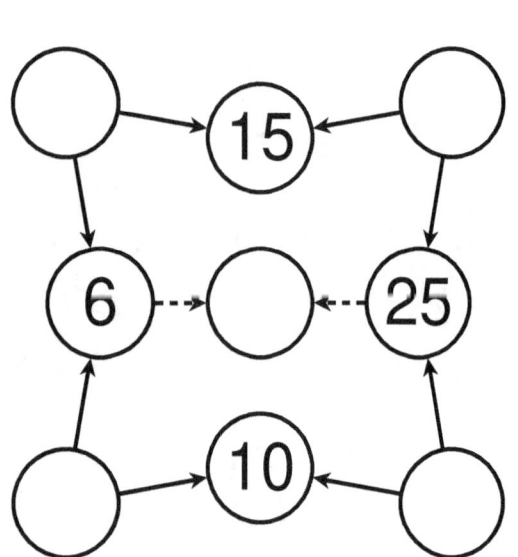

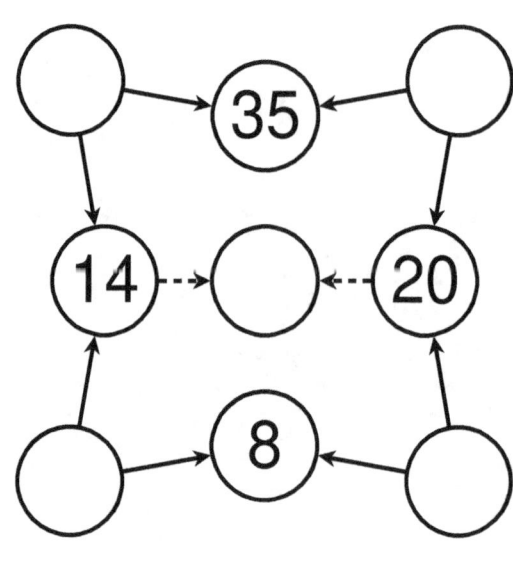

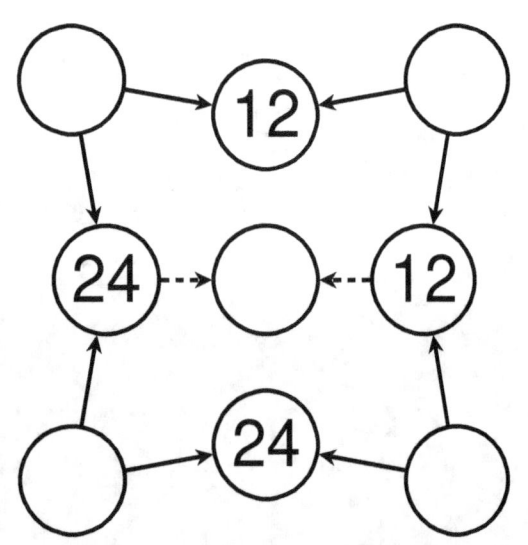

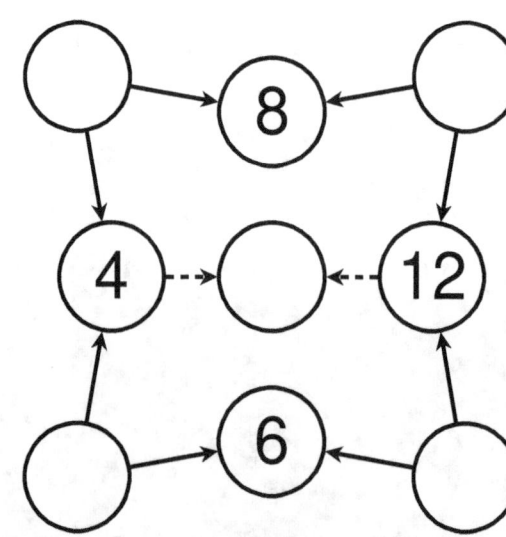

Name _____ Date _____

Find the missing numbers. Solid lines mean multiply.
Dotted lines mean add.

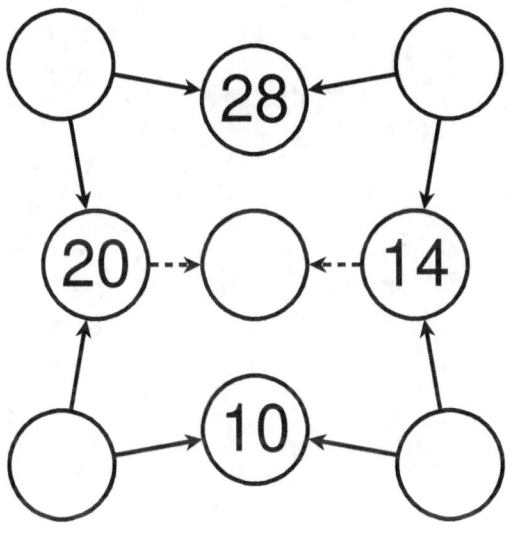

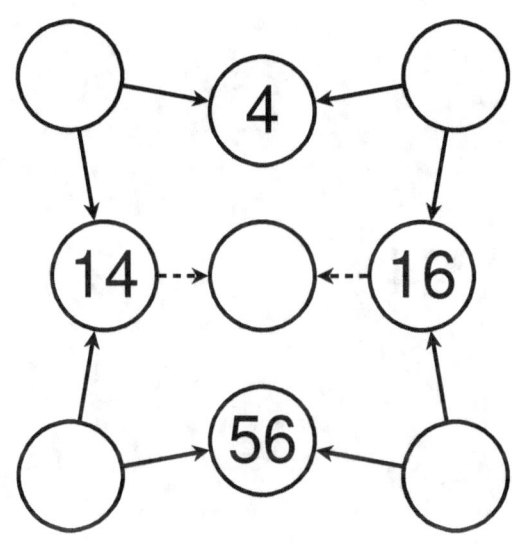

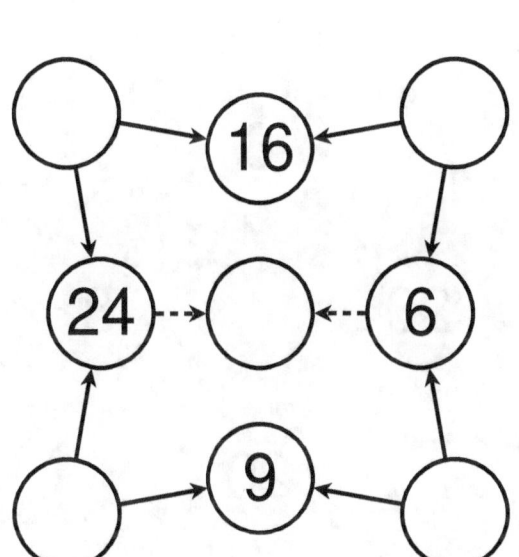

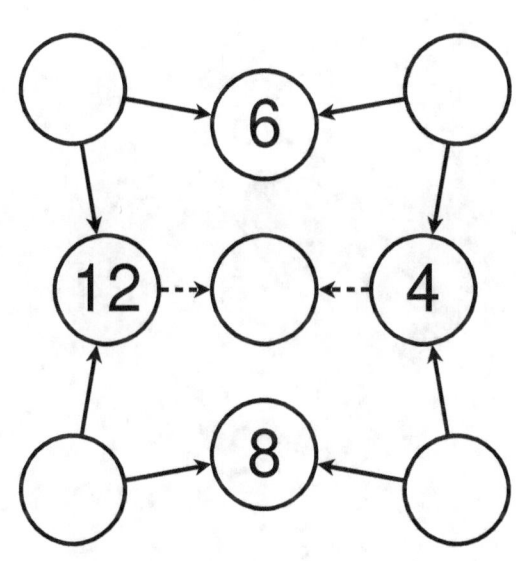

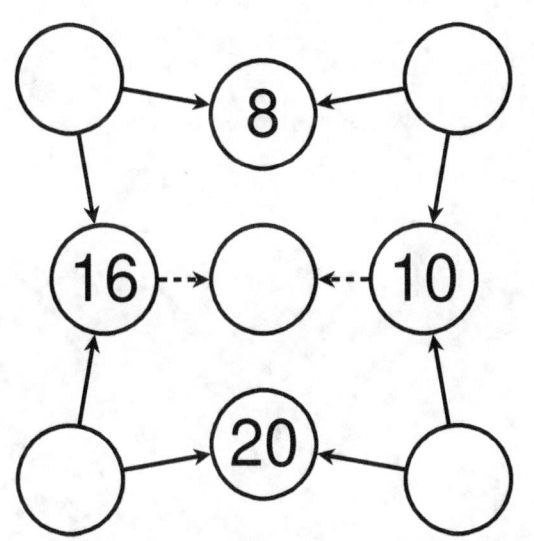

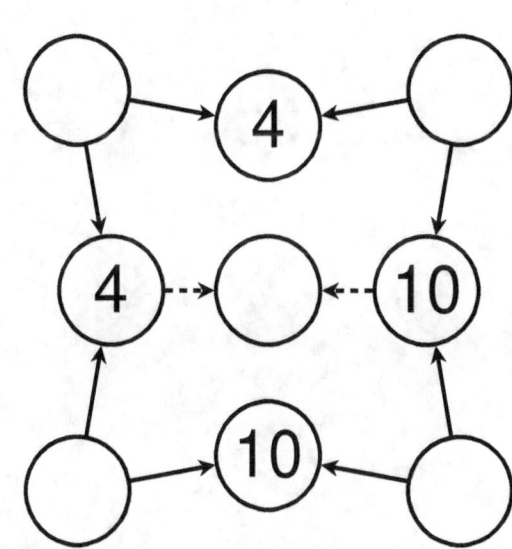

Page 39

Name _____ Date _____

Find the missing numbers. Solid lines mean multiply.
Dotted lines mean add.

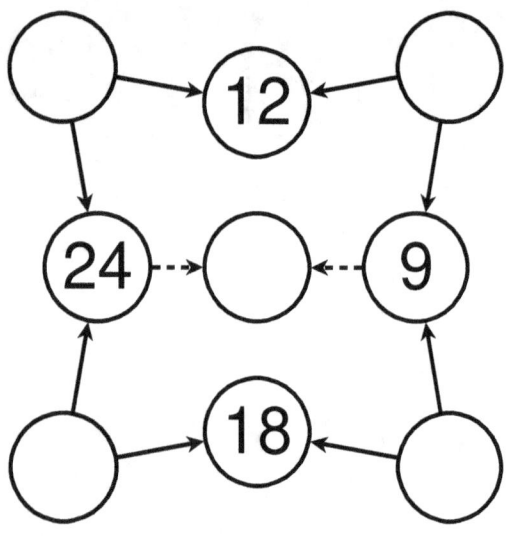

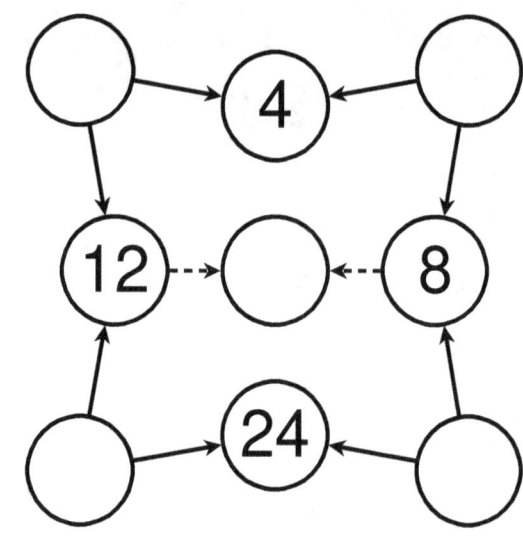

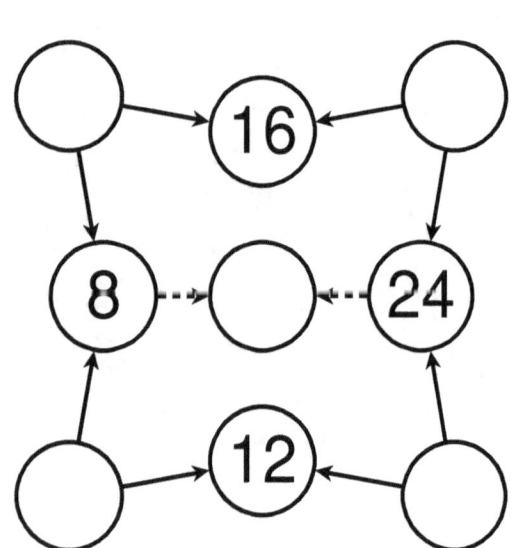

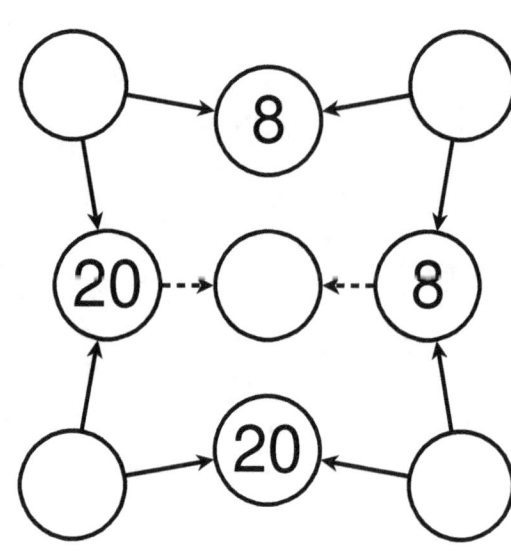

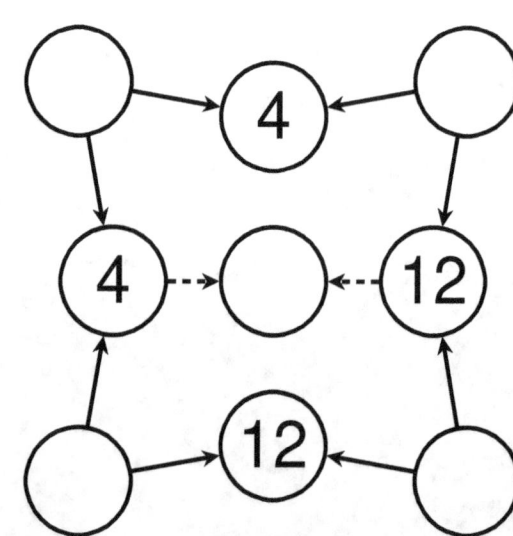

Name _____ Date _____

Find the missing numbers. Solid lines mean multiply.
Dotted lines mean add.

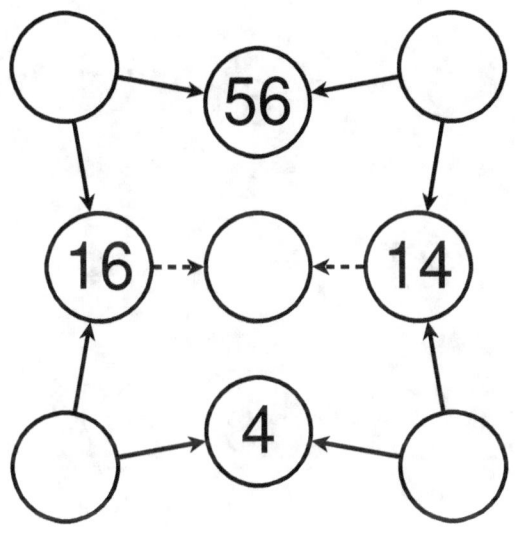
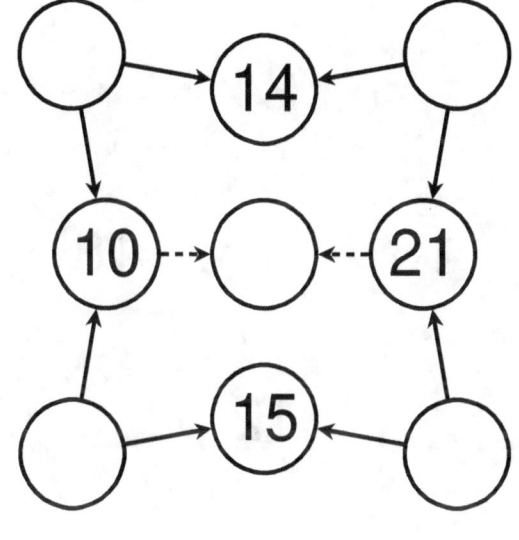

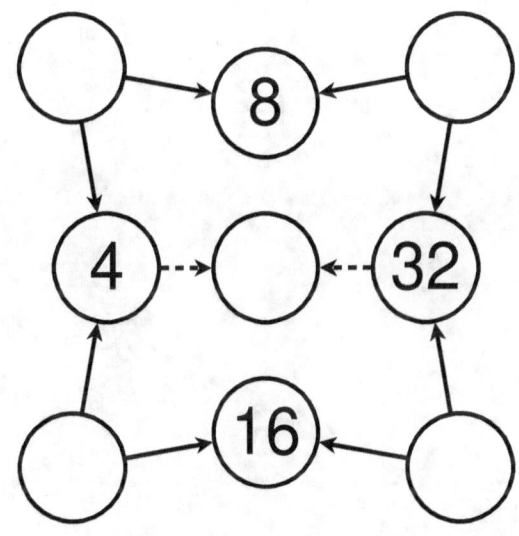
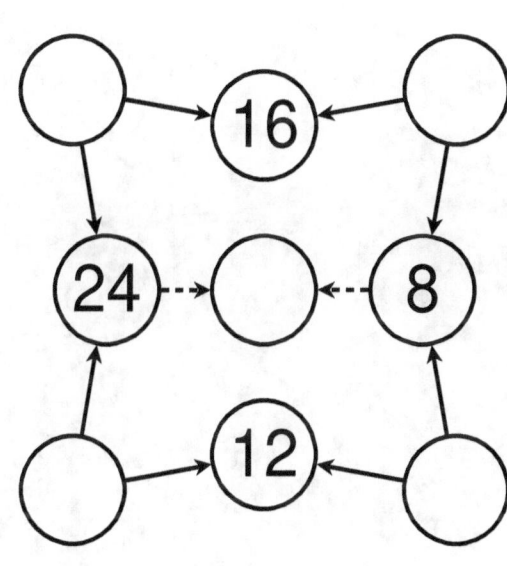

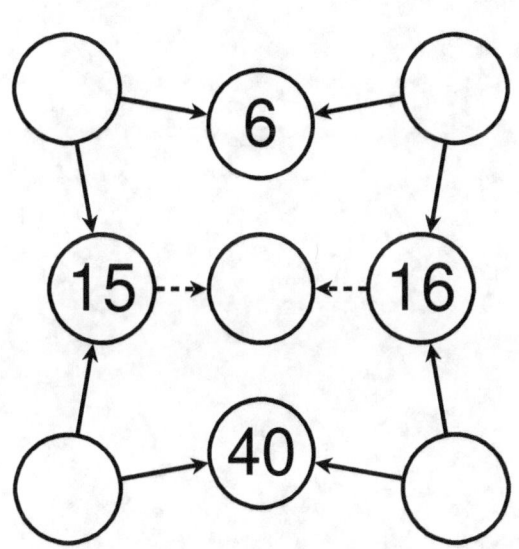
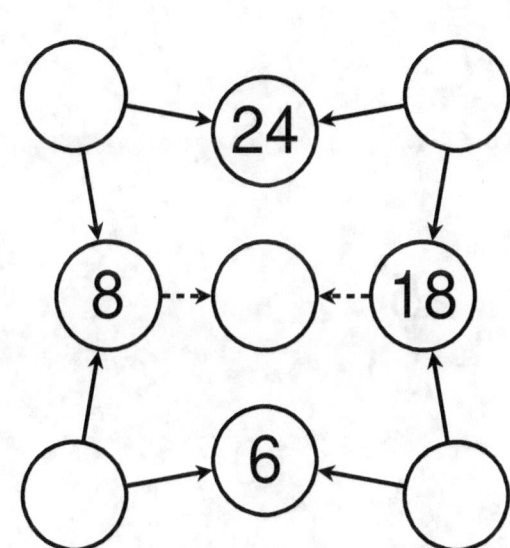

Name _____ Date _____

Find the missing numbers. Solid lines mean multiply.
Dotted lines mean add.

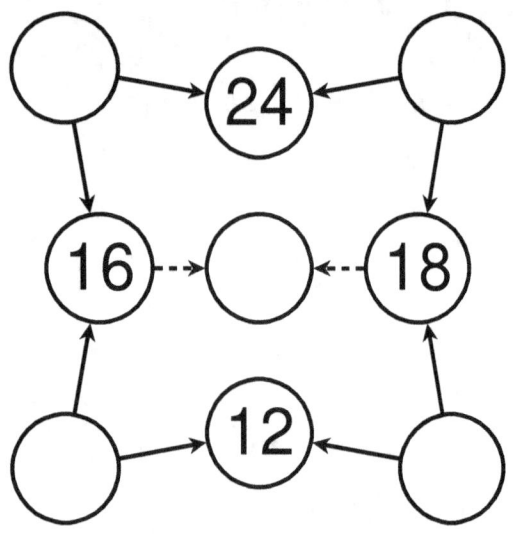
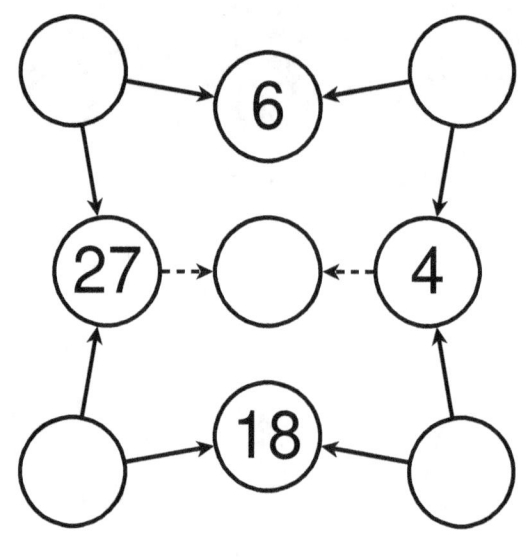
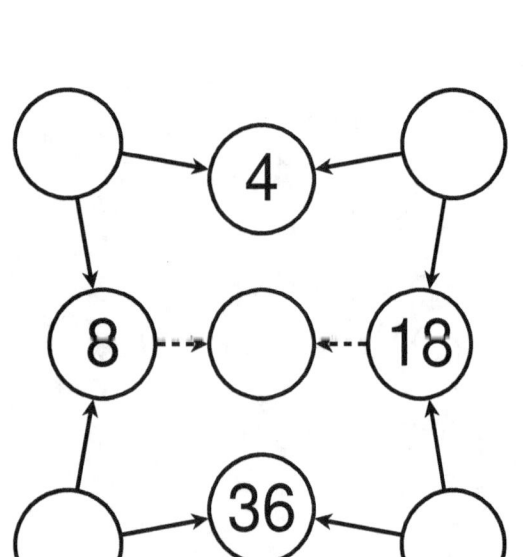
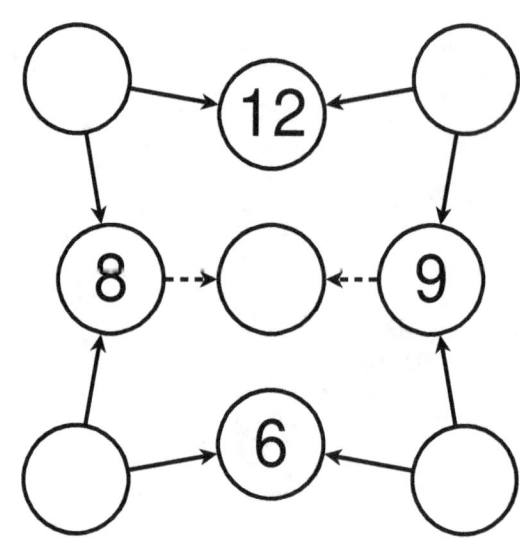
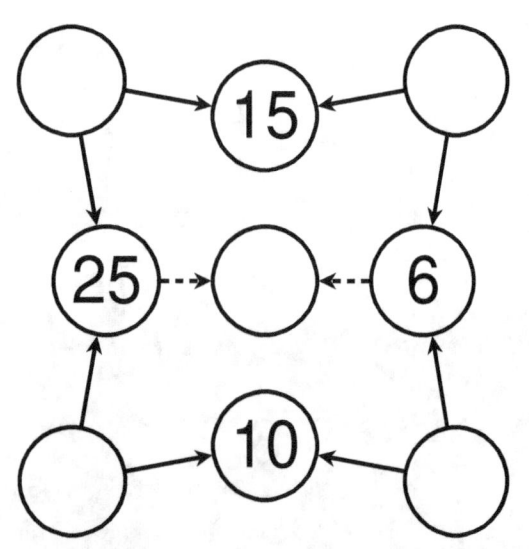
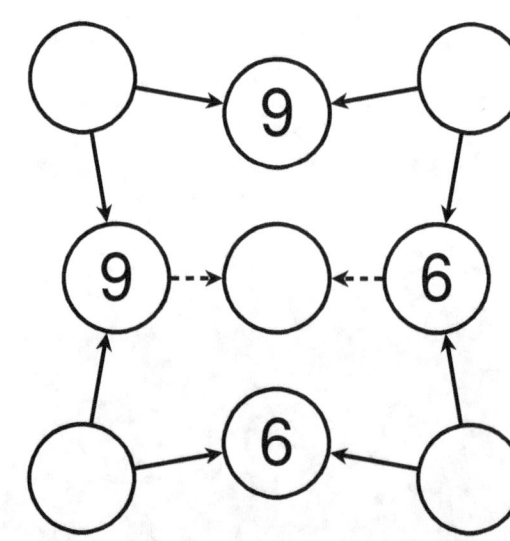

Name _____ Date _____

Find the missing numbers. Solid lines mean multiply.
Dotted lines mean add.

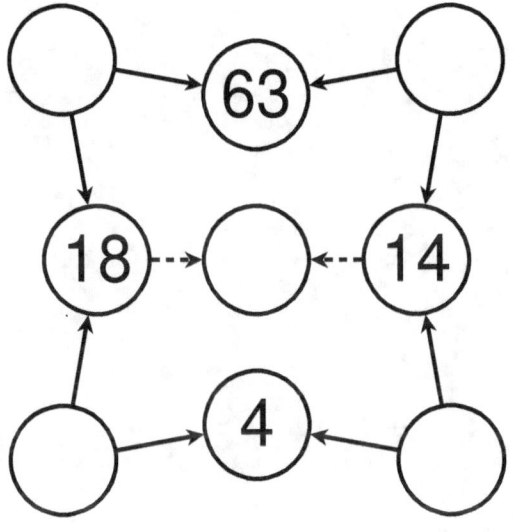

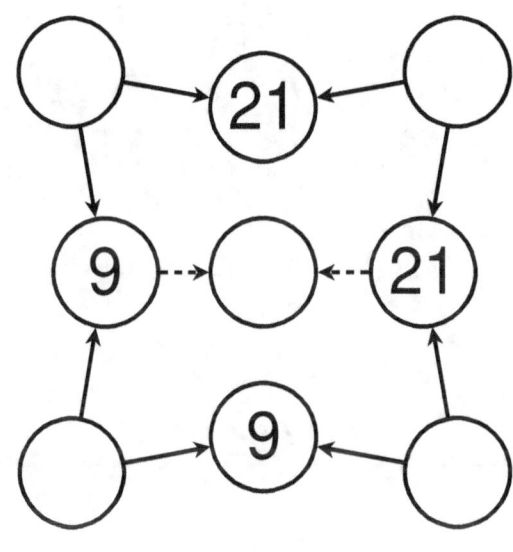

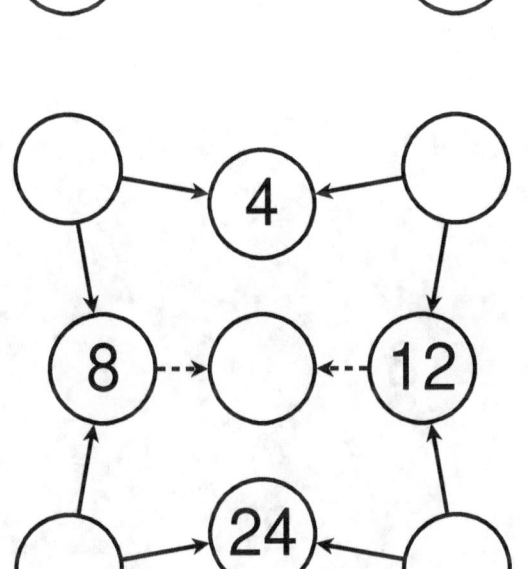

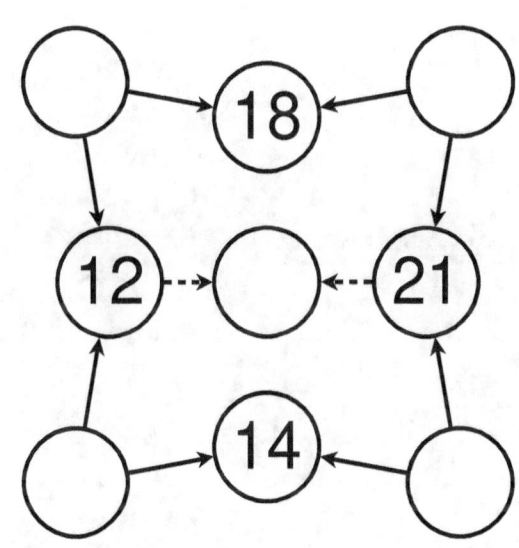

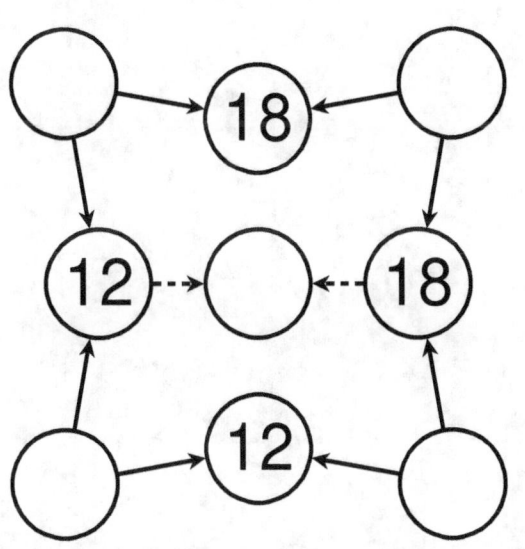

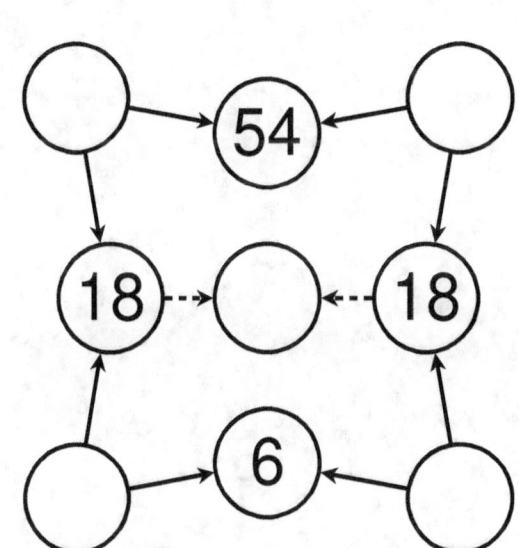

Name _____ Date _____

Find the missing numbers. Solid lines mean multiply.
Dotted lines mean add.

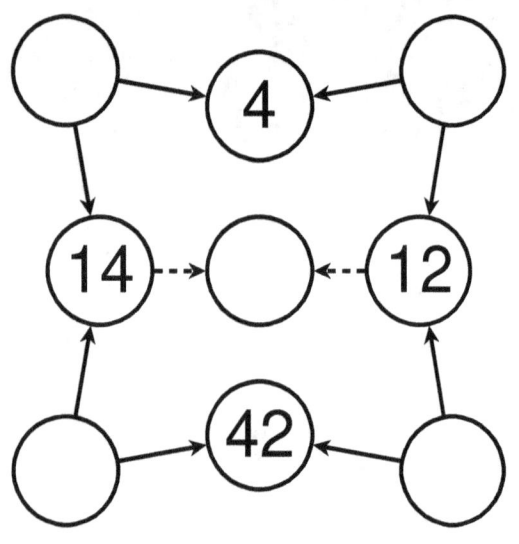

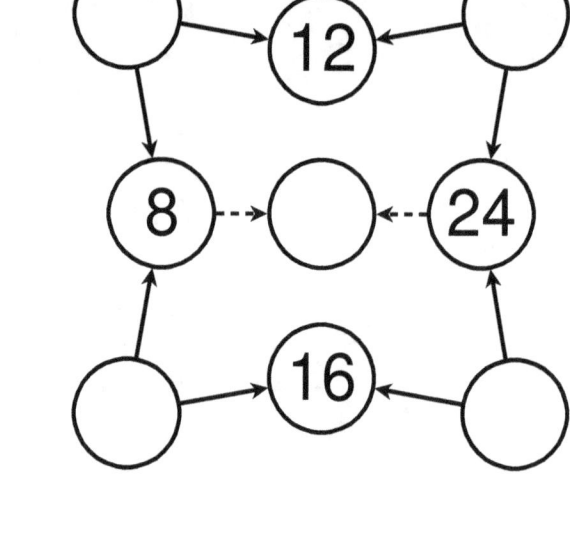

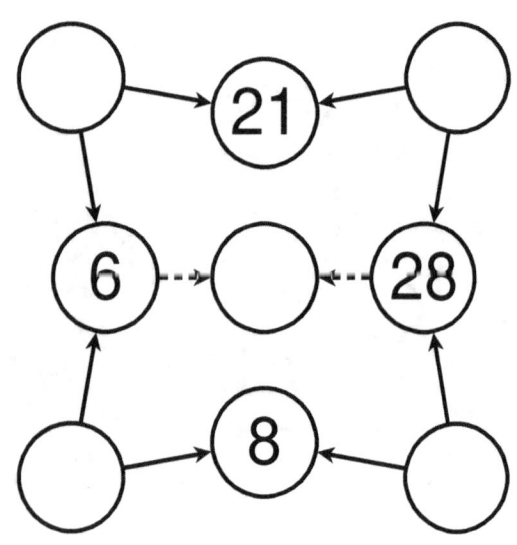

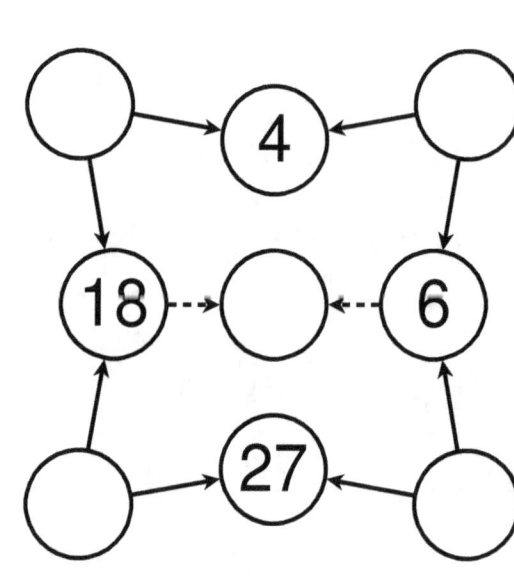

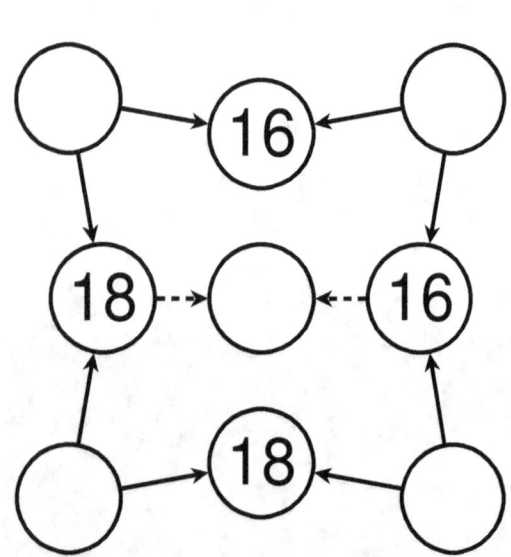

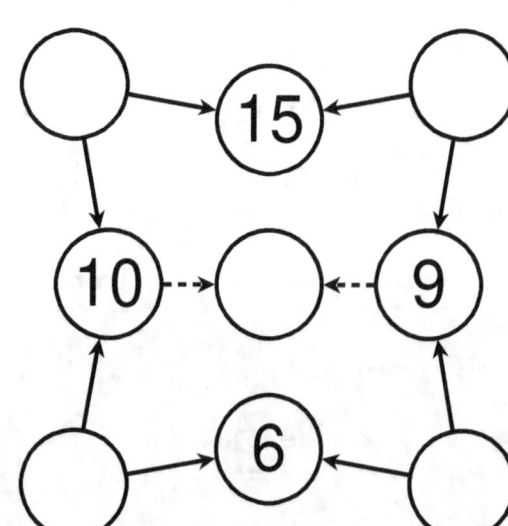

Name _____ Date _____

Find the missing numbers. Solid lines mean multiply.
Dotted lines mean add.

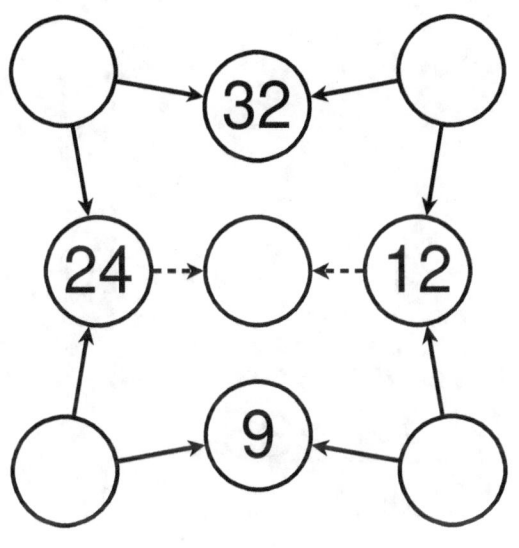

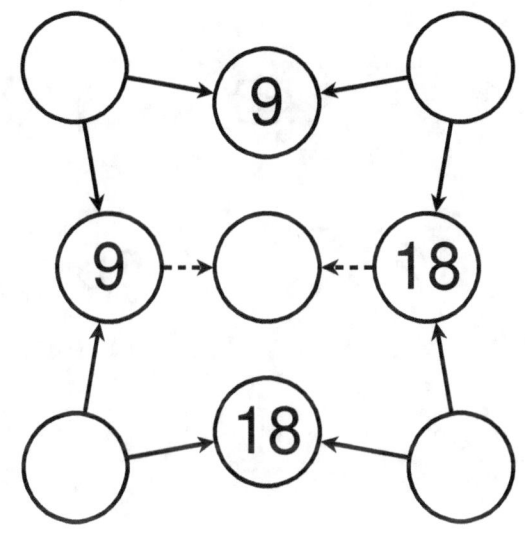

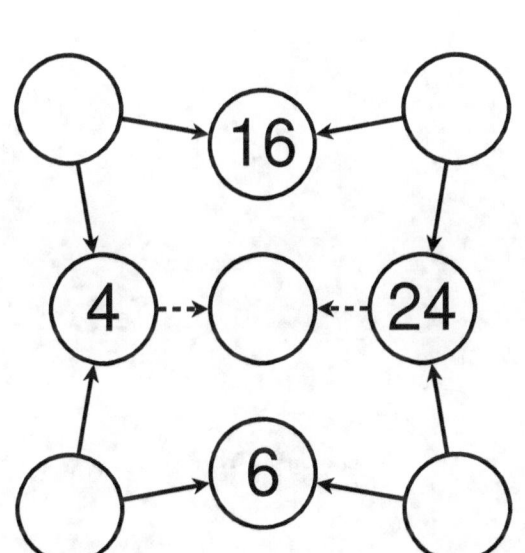

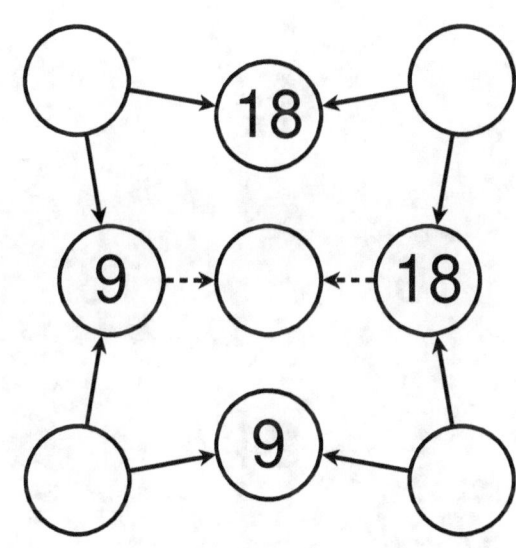

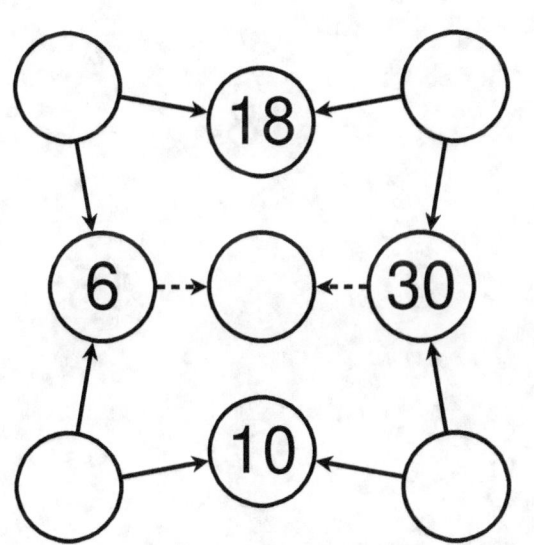

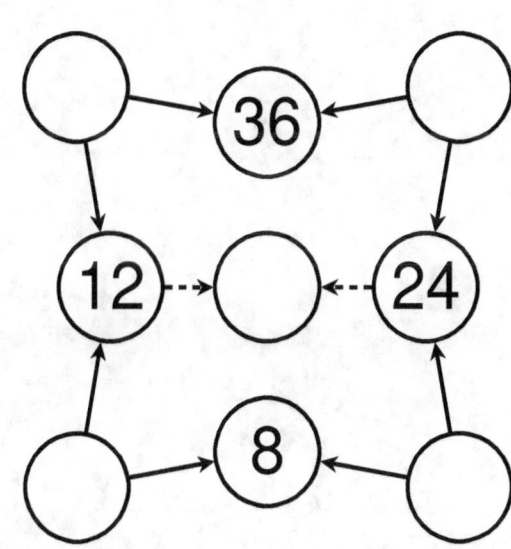

Name _____ Date _____

Find the missing numbers. Solid lines mean multiply.
Dotted lines mean add.

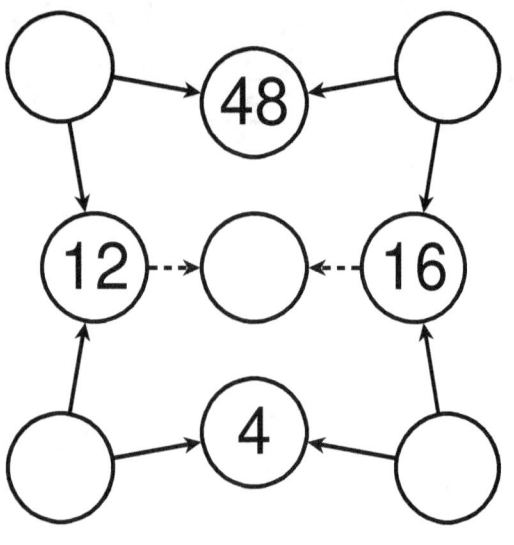

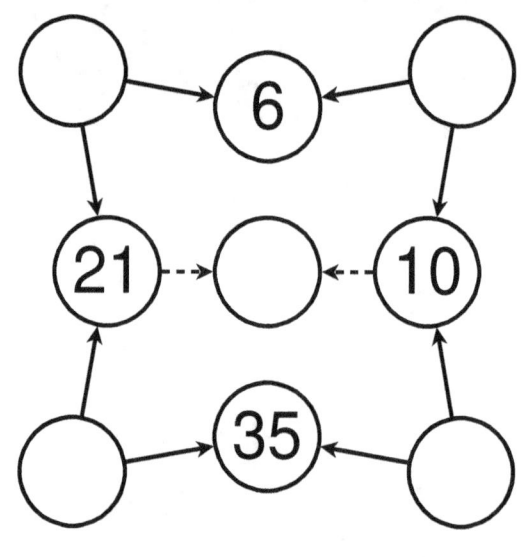

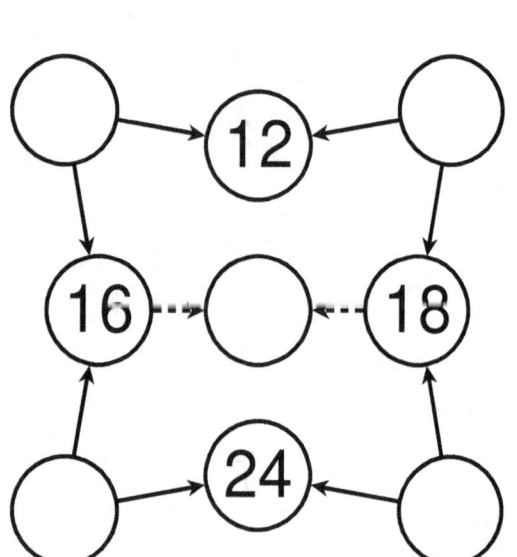

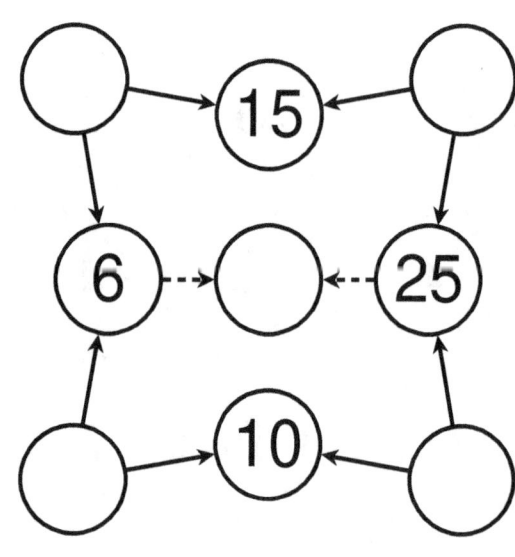

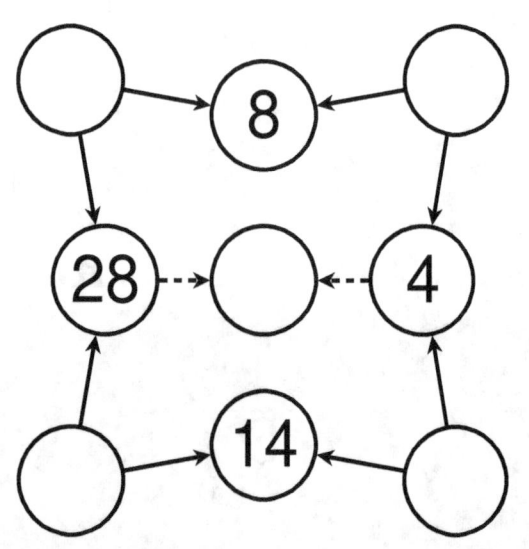

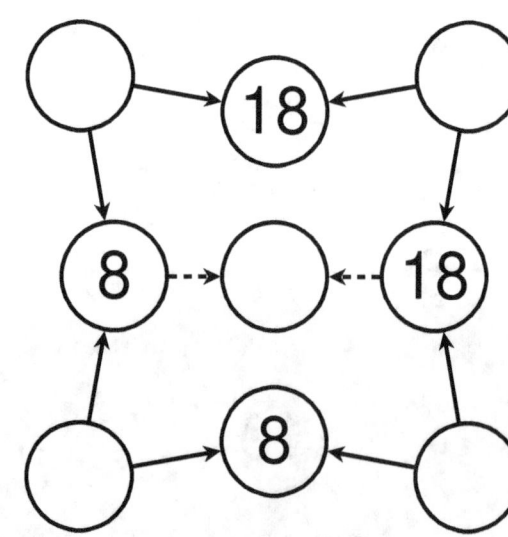

Name _____ Date _____

Find the missing numbers. Solid lines mean multiply.
Dotted lines mean add.

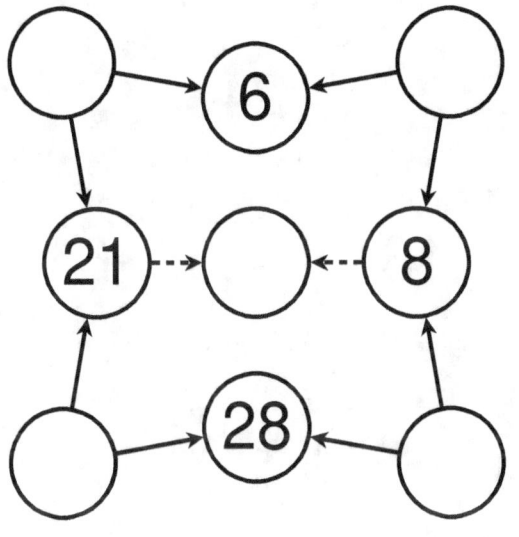

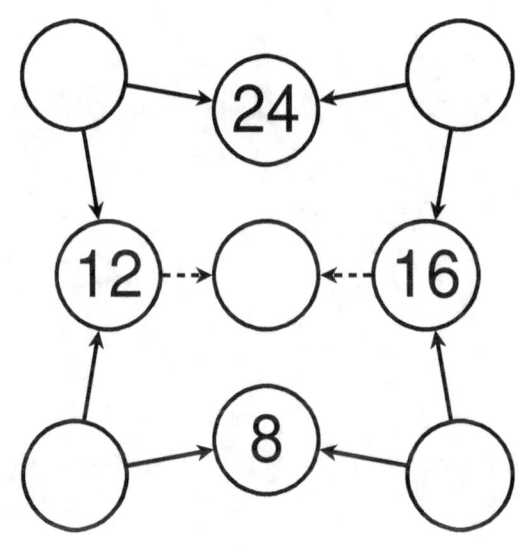

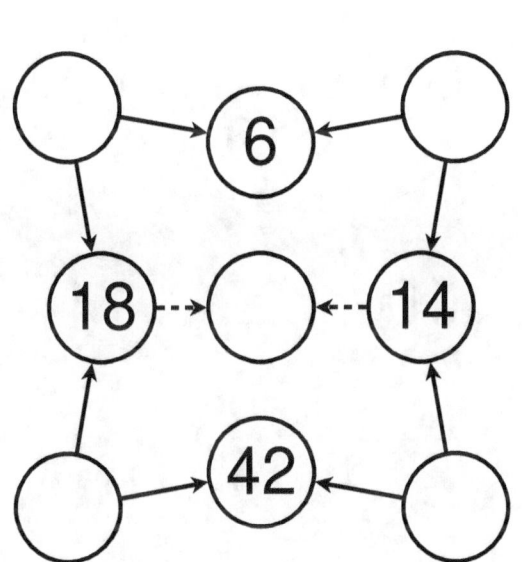

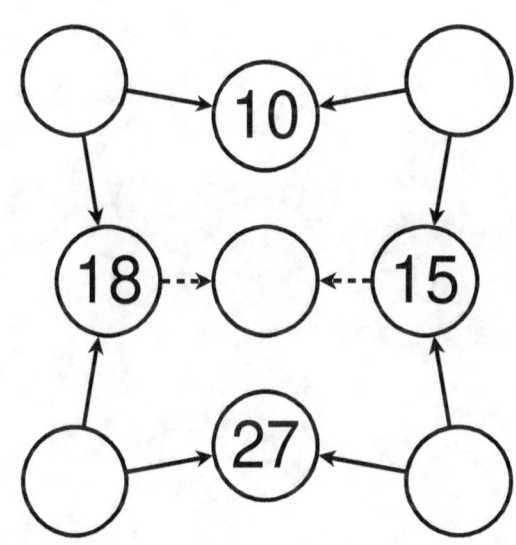

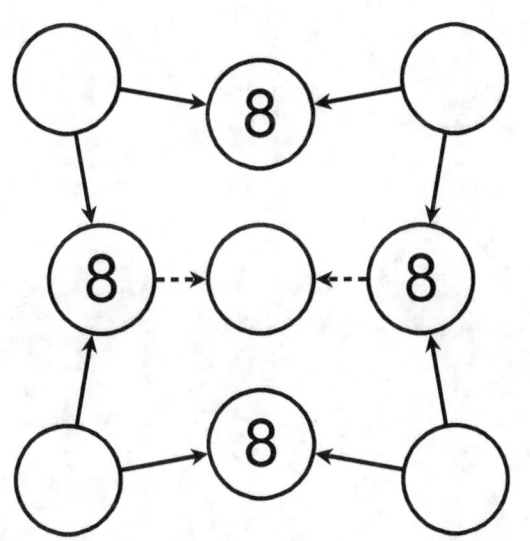

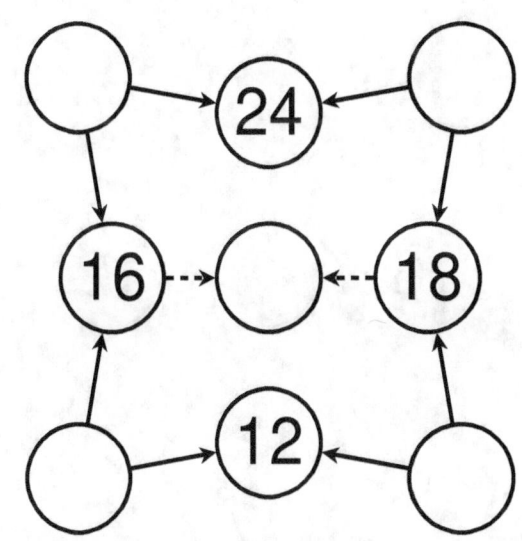

Name _____ Date _____

Find the missing numbers. Solid lines mean multiply.
Dotted lines mean add.

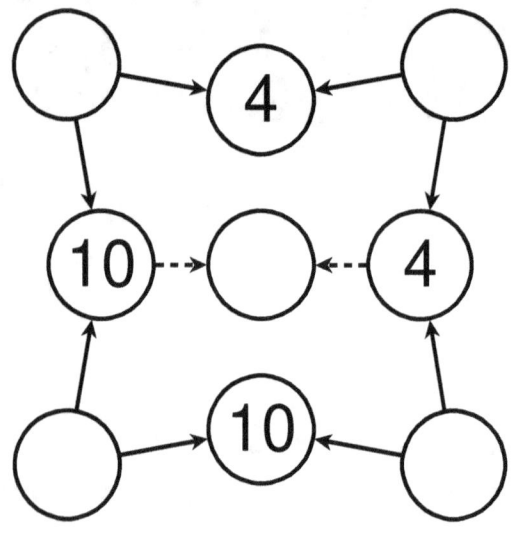

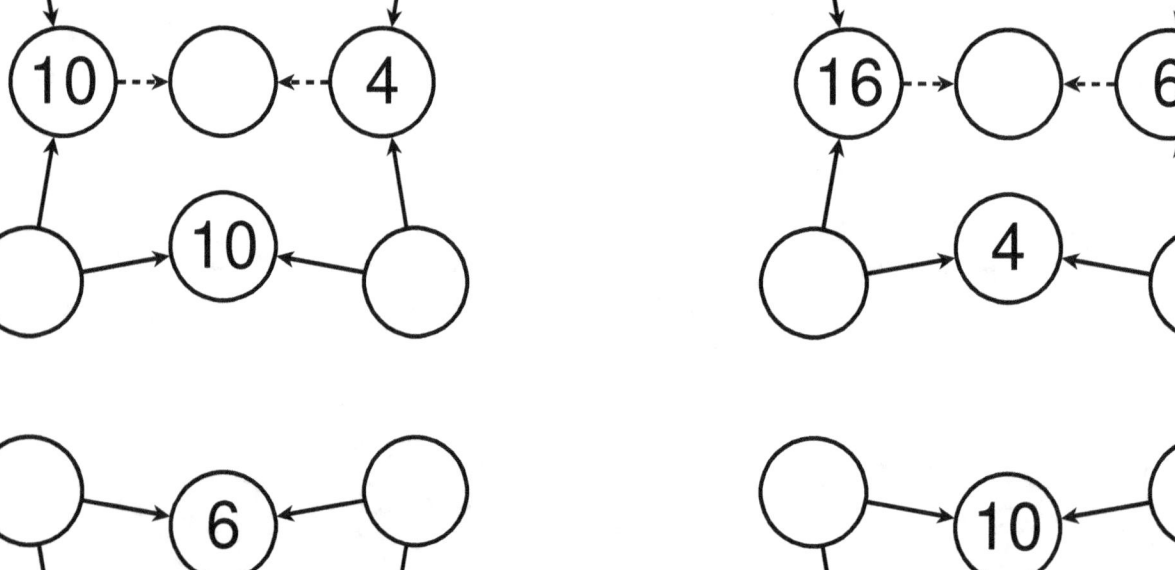

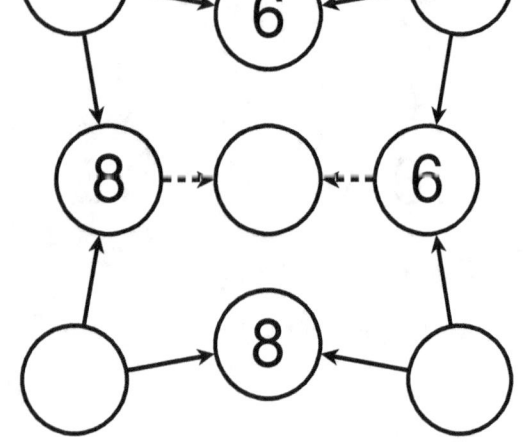

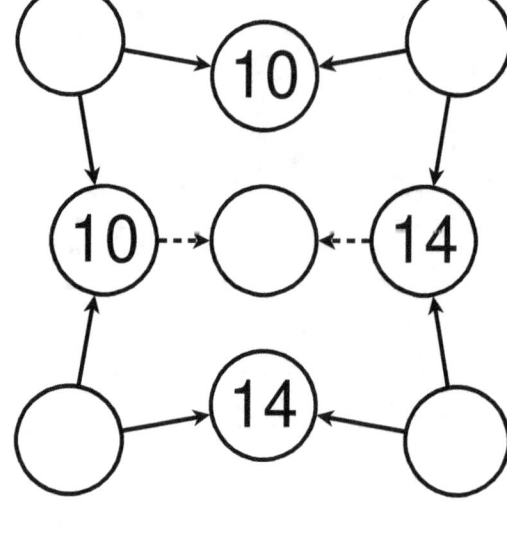

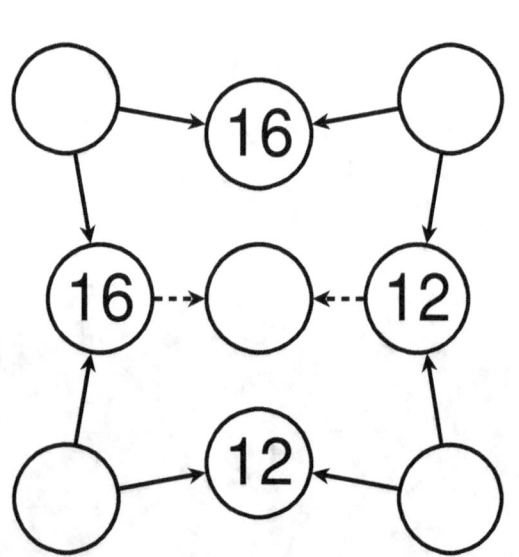

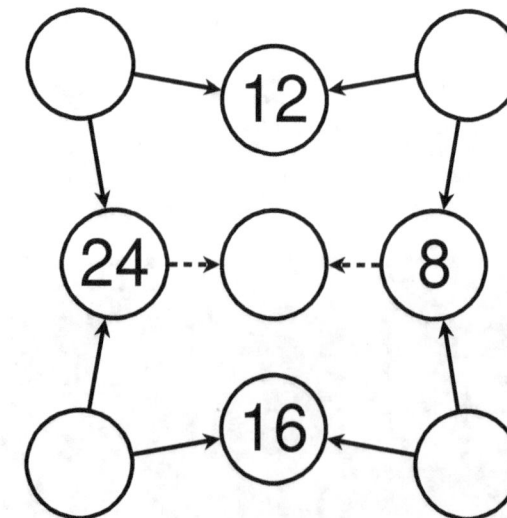

Name _____ Date _____

Find the missing numbers. Solid lines mean multiply.
Dotted lines mean add.

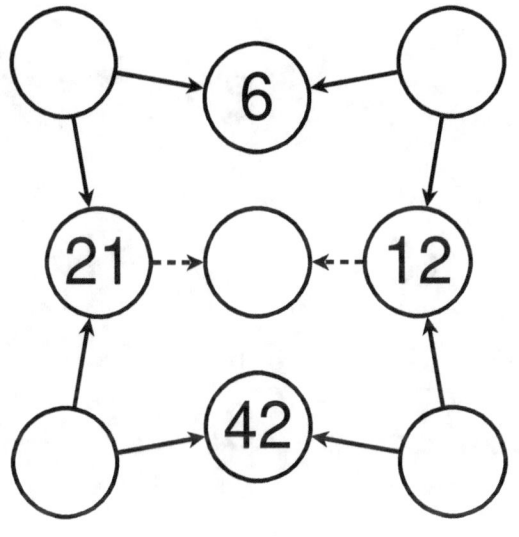

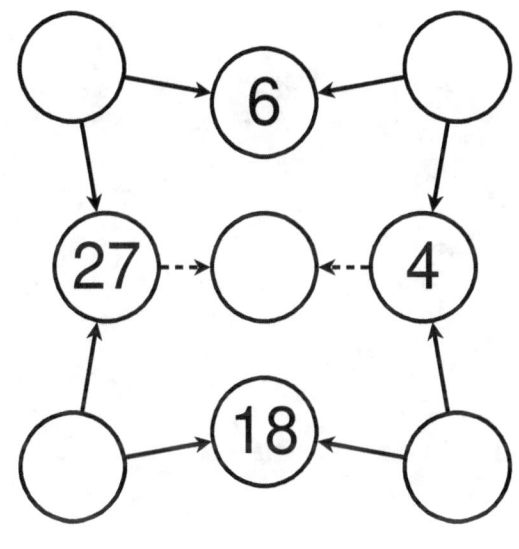

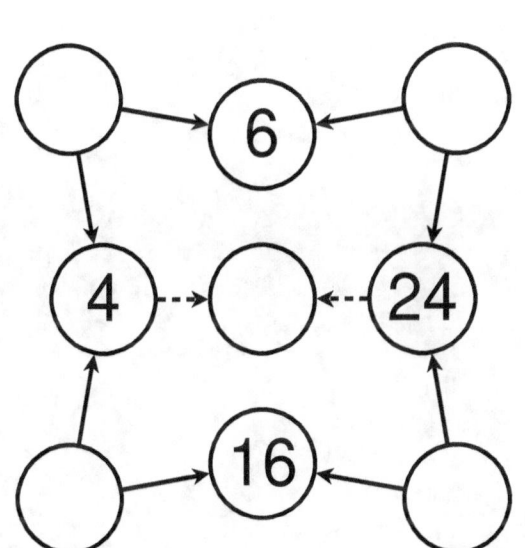

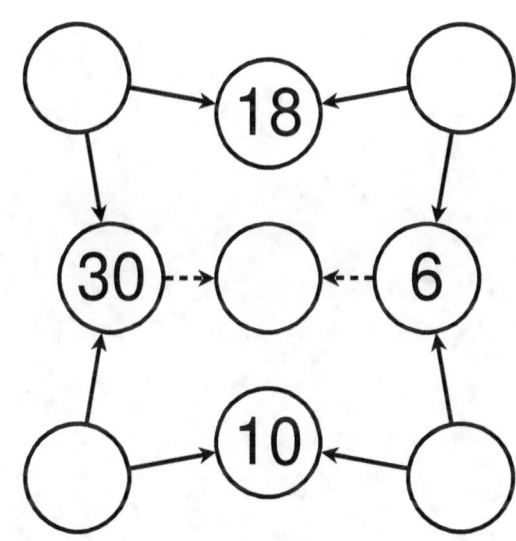

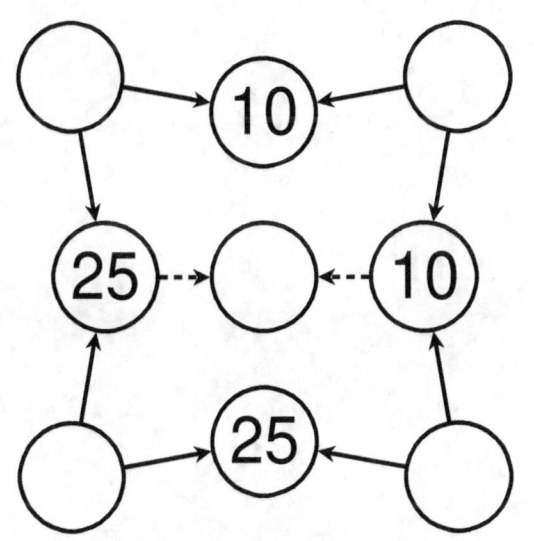

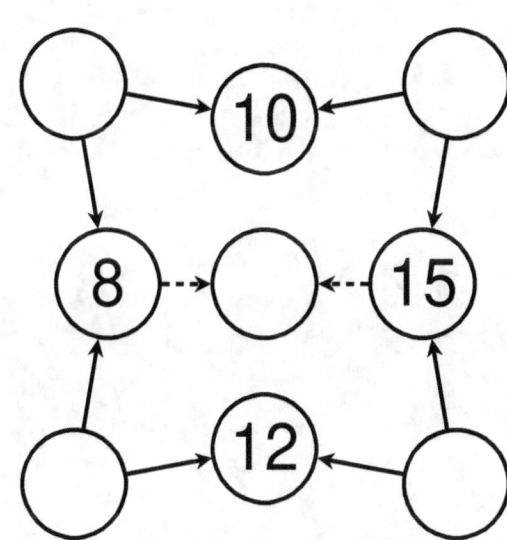

Name _____ Date _____

Find the missing numbers. Solid lines mean multiply.
Dotted lines mean add.

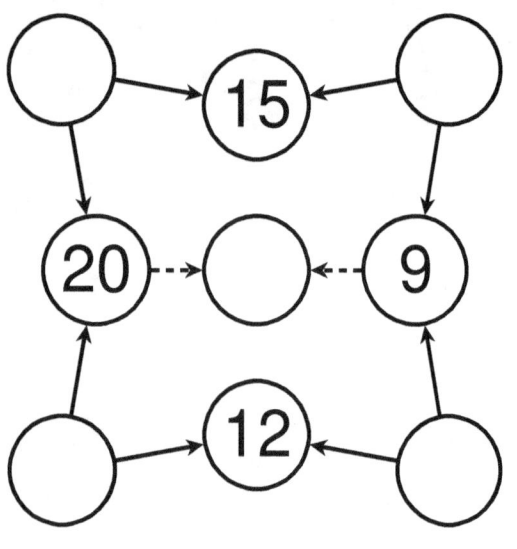

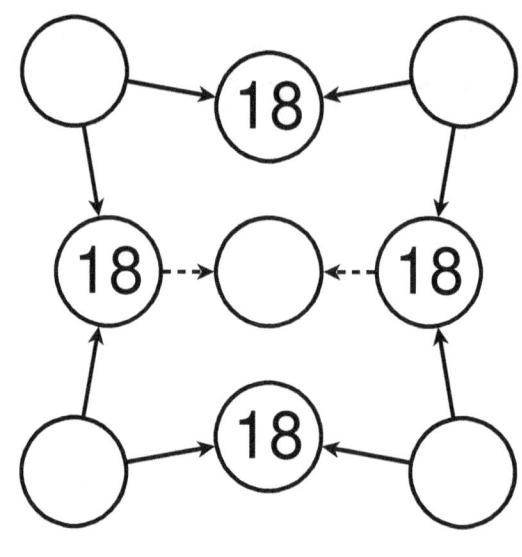

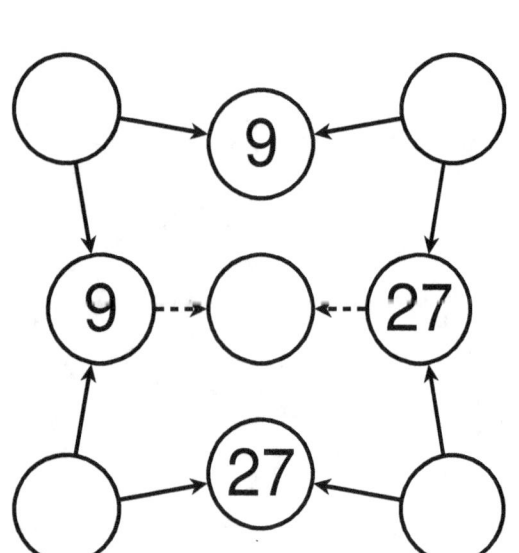

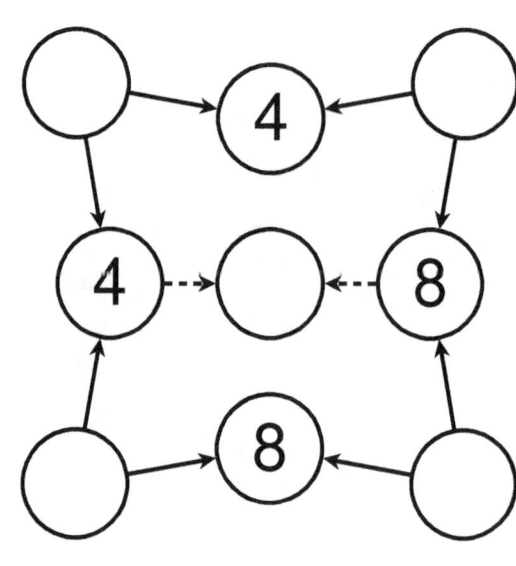

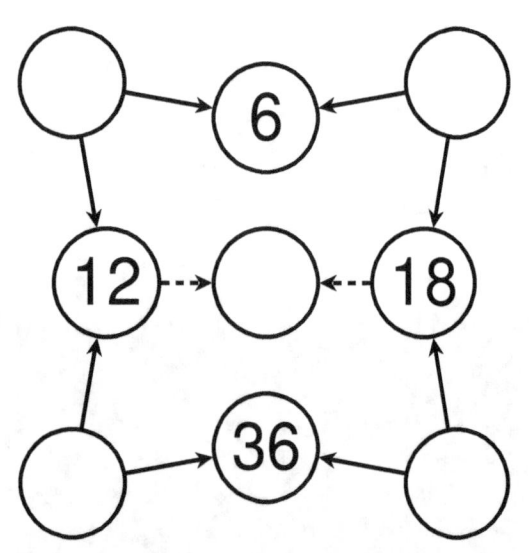

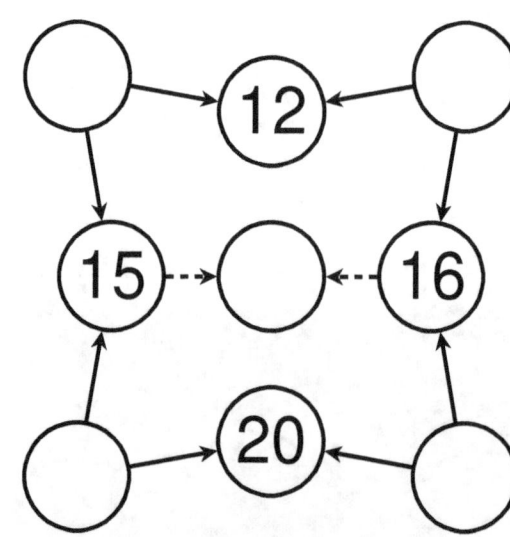

Name _____ Date _____

Find the missing numbers. Solid lines mean multiply.
Dotted lines mean add.

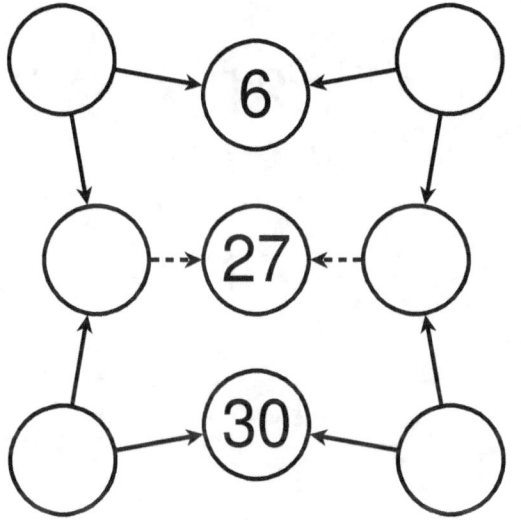

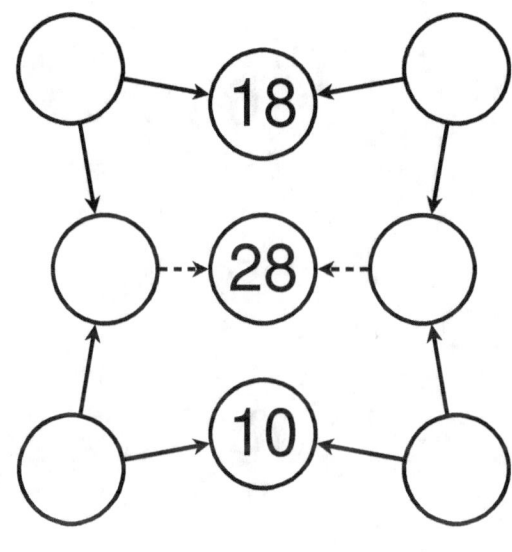

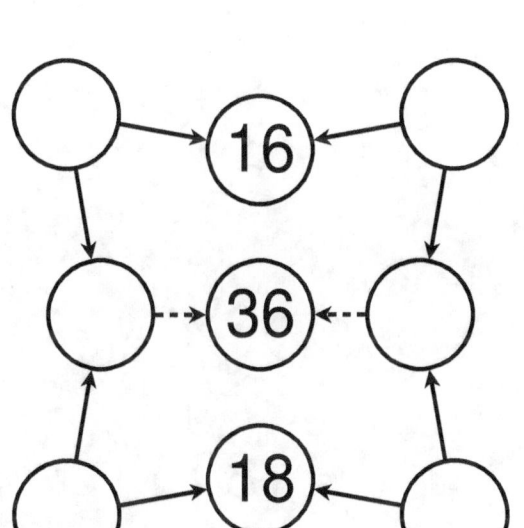

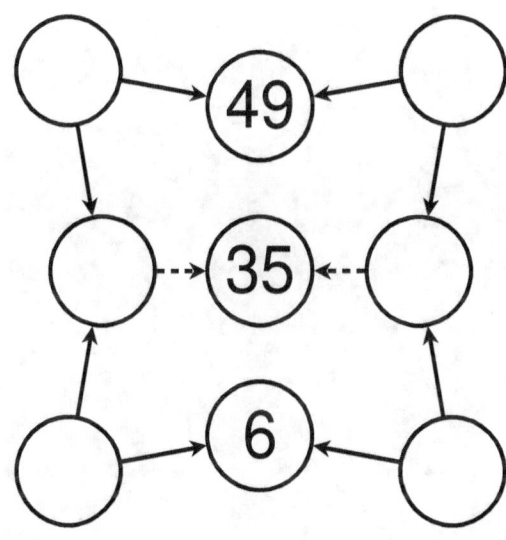

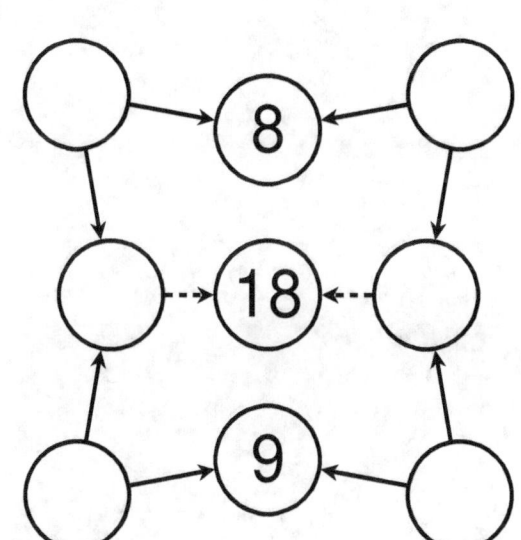

Name _____ Date _____

Find the missing numbers. Solid lines mean multiply.
Dotted lines mean add.

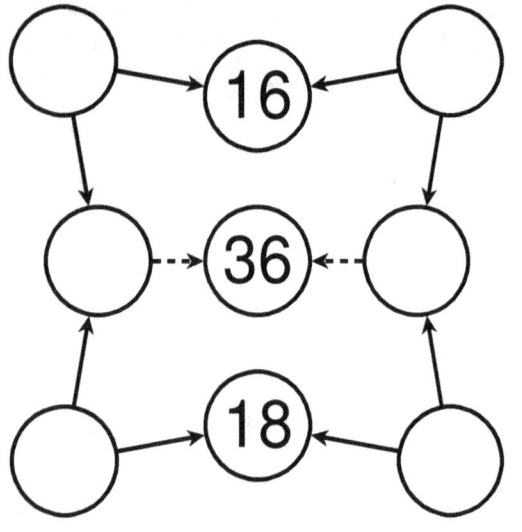

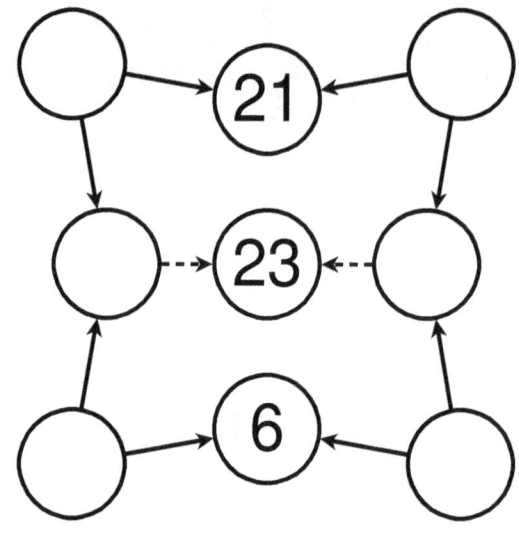

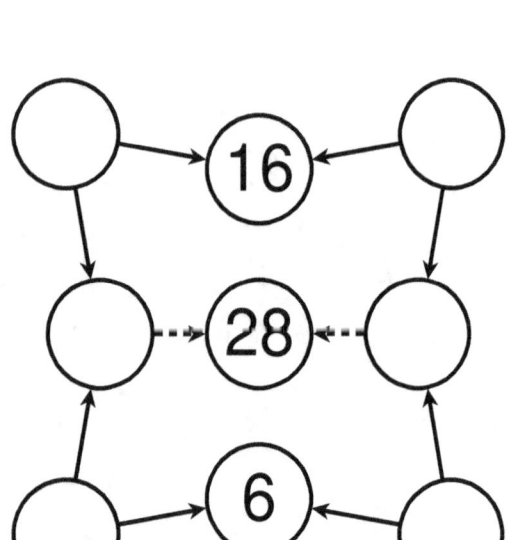

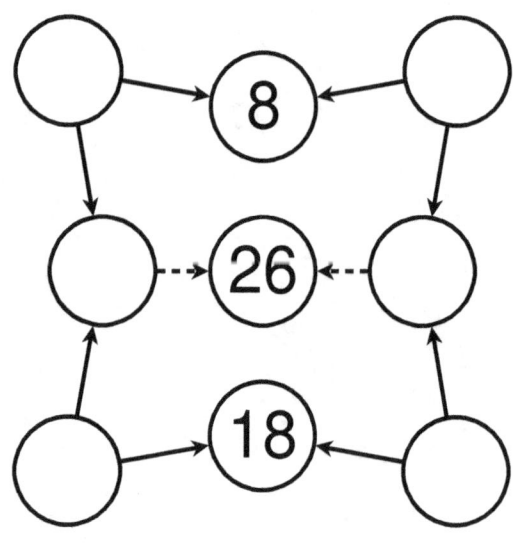

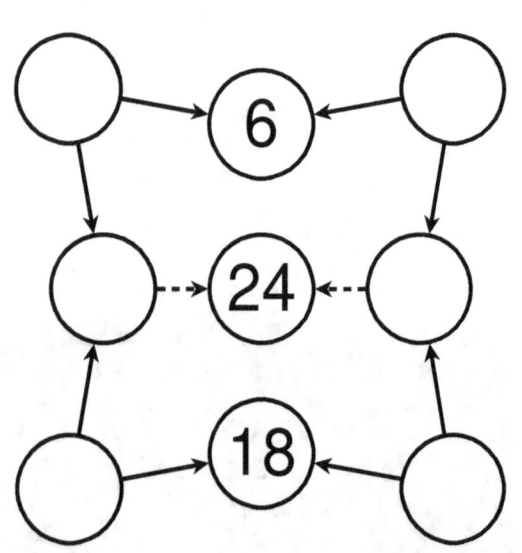

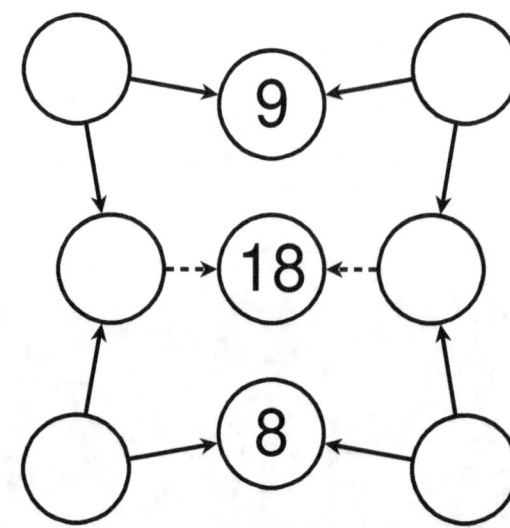

Name _____ Date _____

Find the missing numbers. Solid lines mean multiply.
Dotted lines mean add.

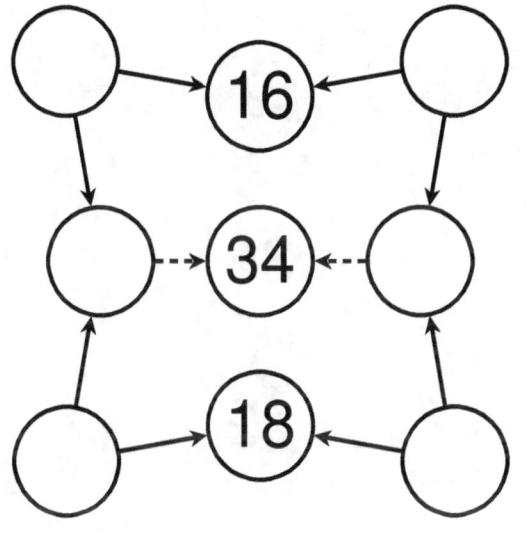

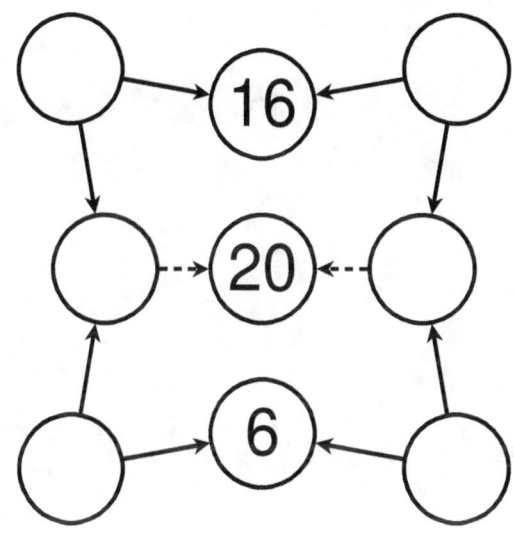

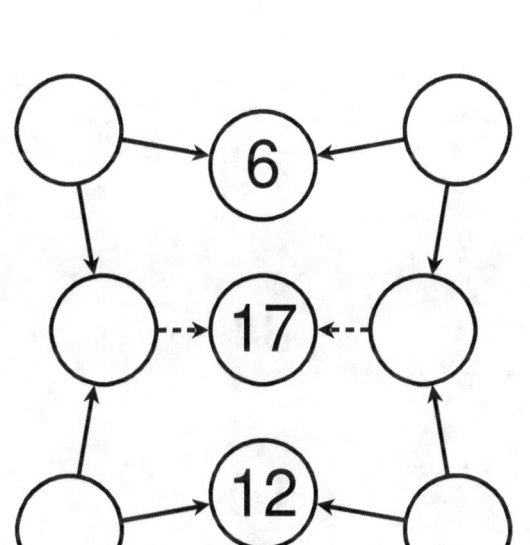

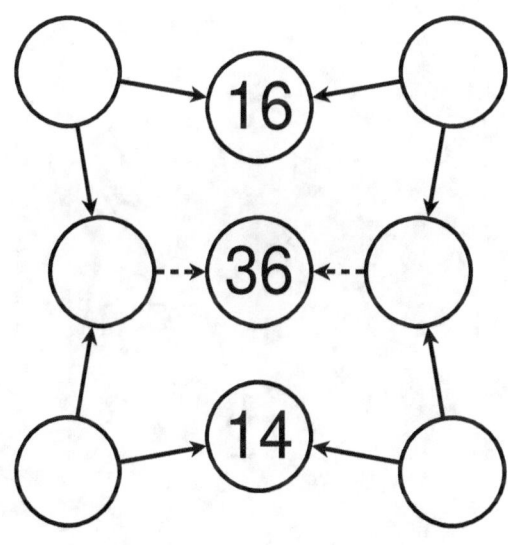

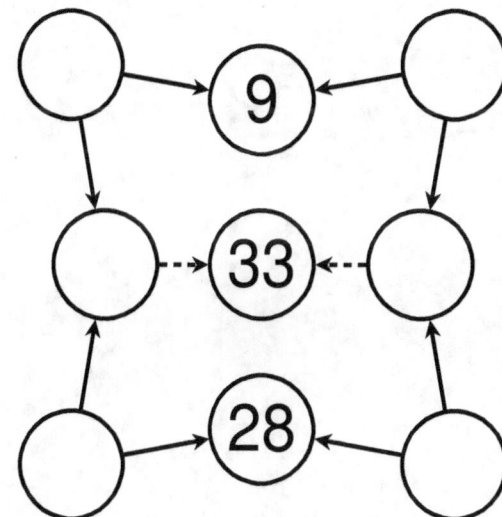

Page 53

Name _____ Date _____

Find the missing numbers. Solid lines mean multiply.
Dotted lines mean add.

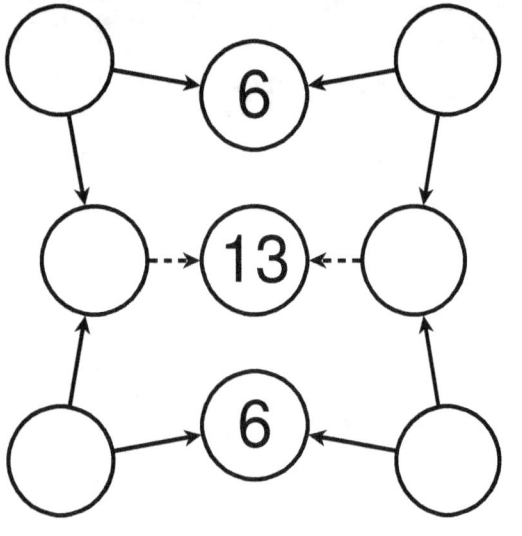

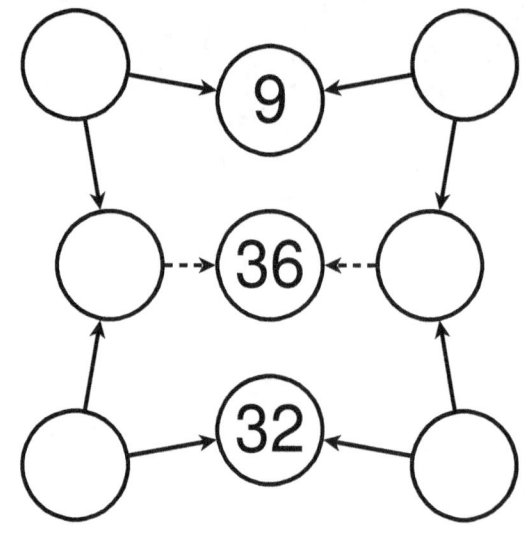

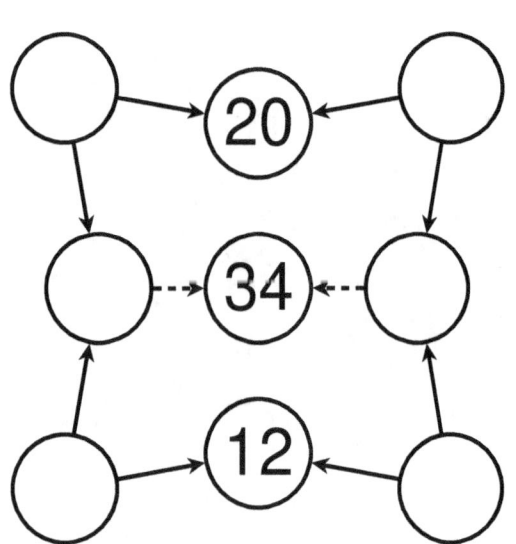

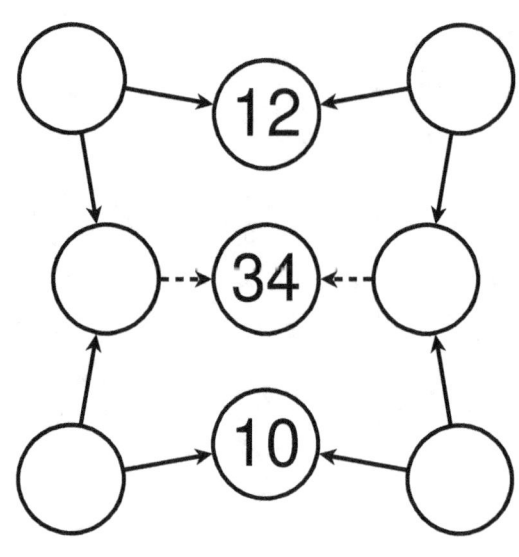

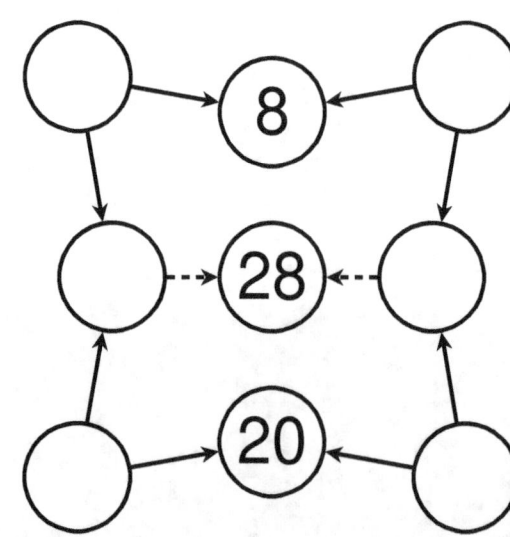

Page 54

Name _____ Date _____

Find the missing numbers. Solid lines mean multiply.
Dotted lines mean add.

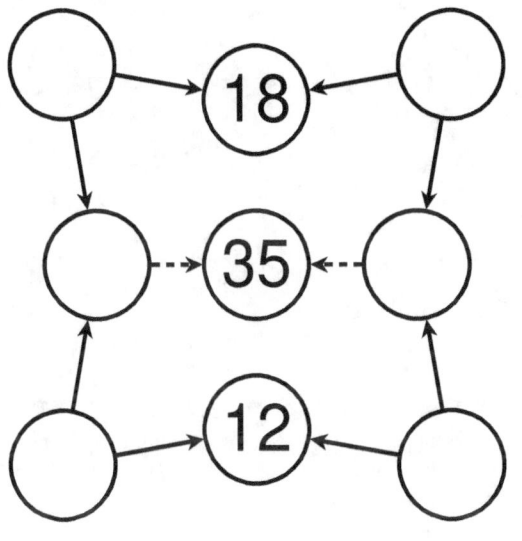

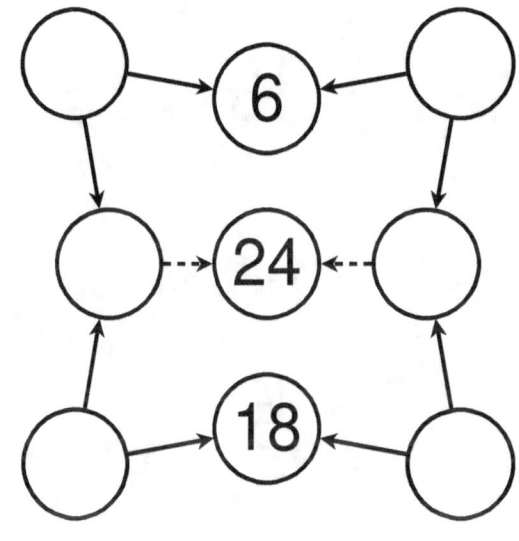

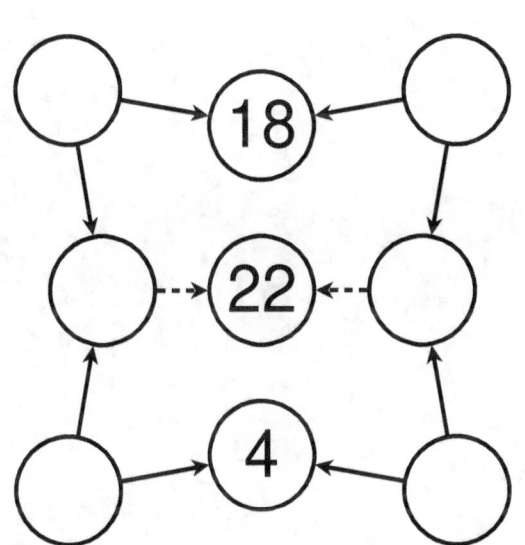

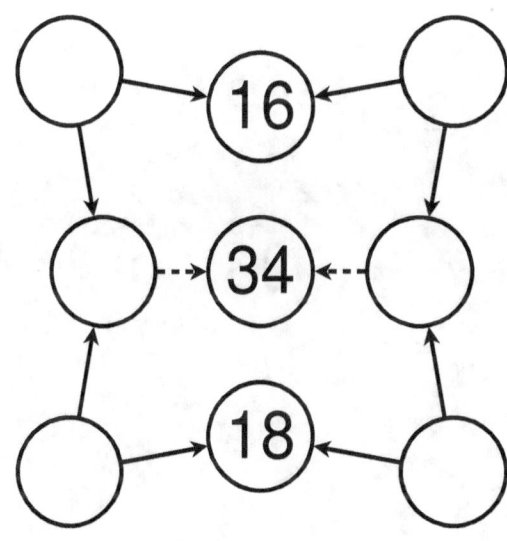

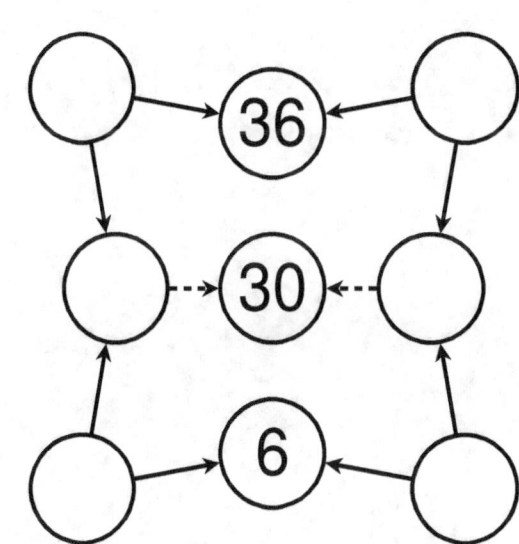

Name _____ Date _____

Find the missing numbers. Solid lines mean multiply.
Dotted lines mean add.

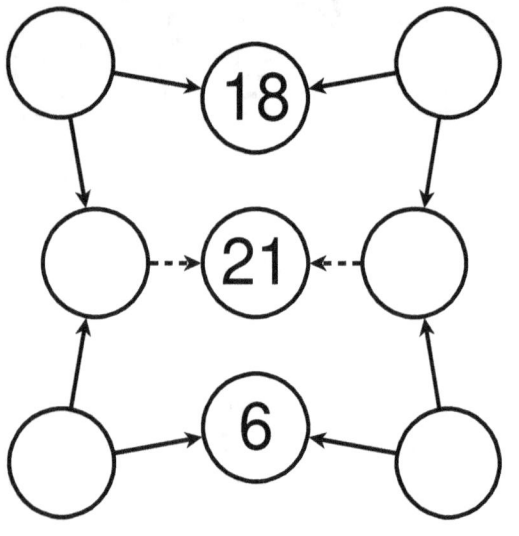

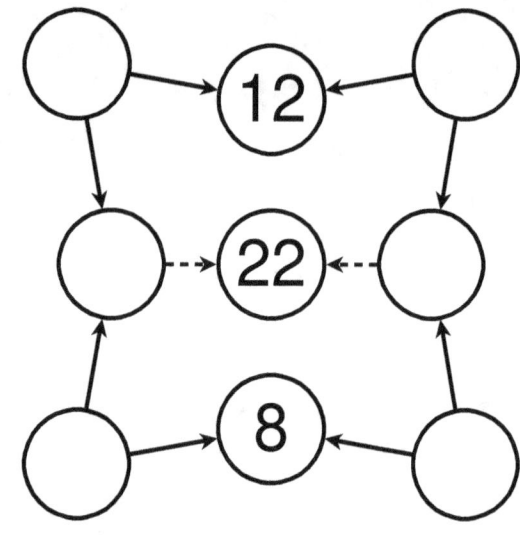

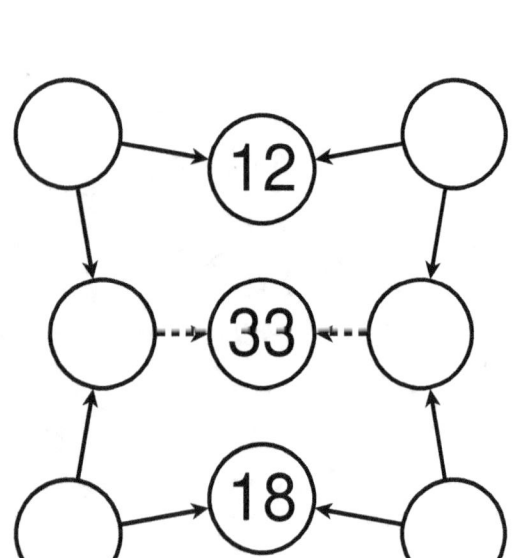

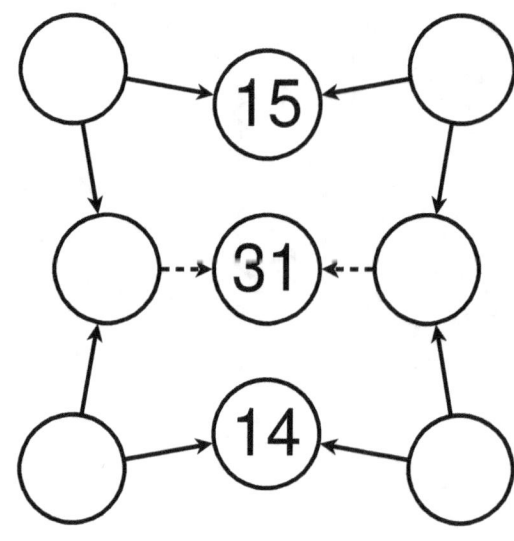

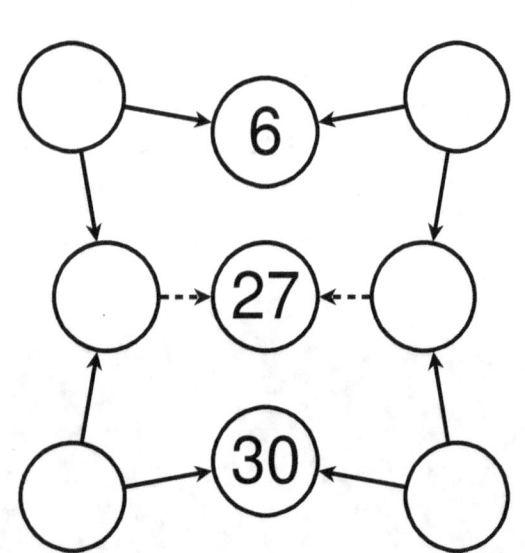

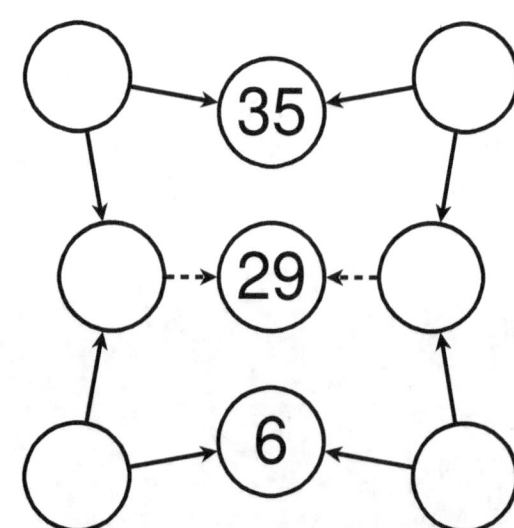

Page 56

Name _____ Date _____

Find the missing numbers. Solid lines mean multiply.
Dotted lines mean add.

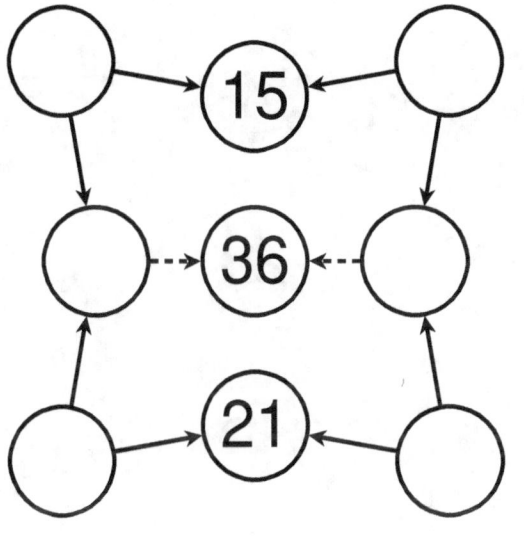

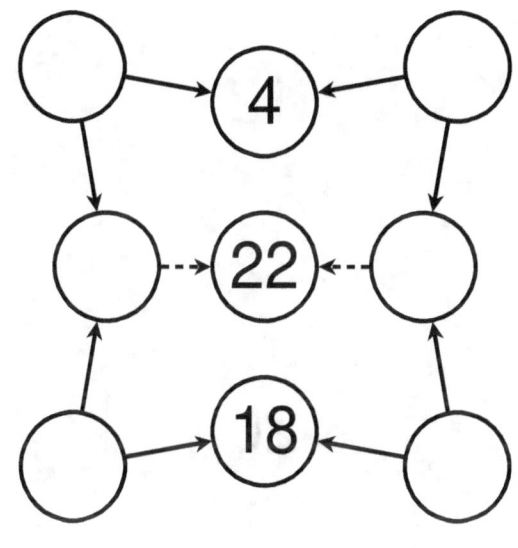

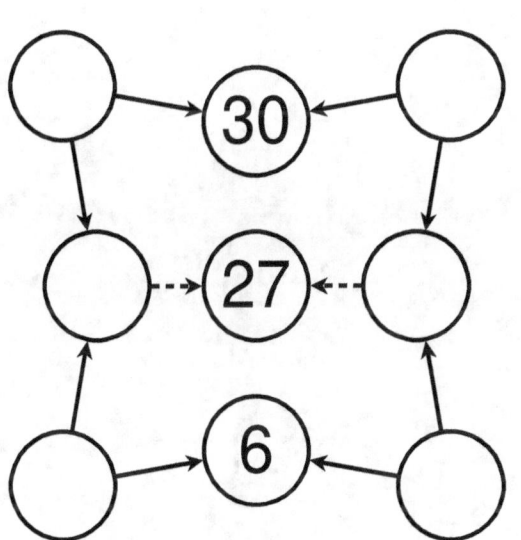

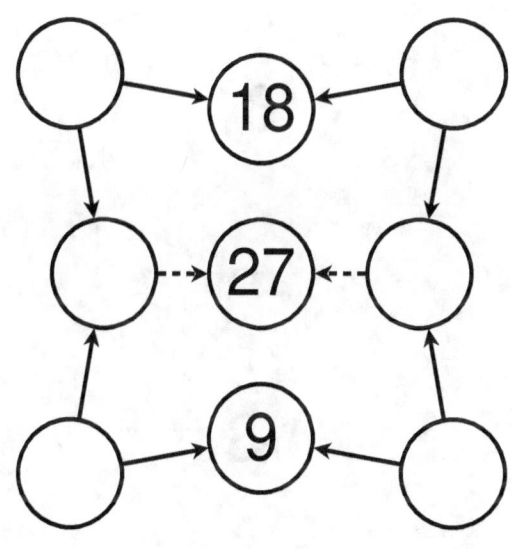

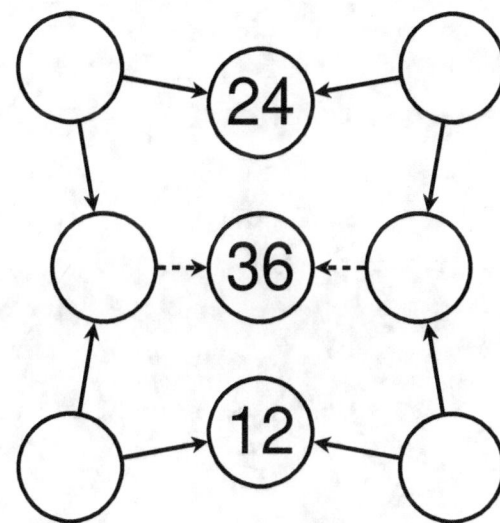

Name _____ Date _____

Find the missing numbers. Solid lines mean multiply.
Dotted lines mean add.

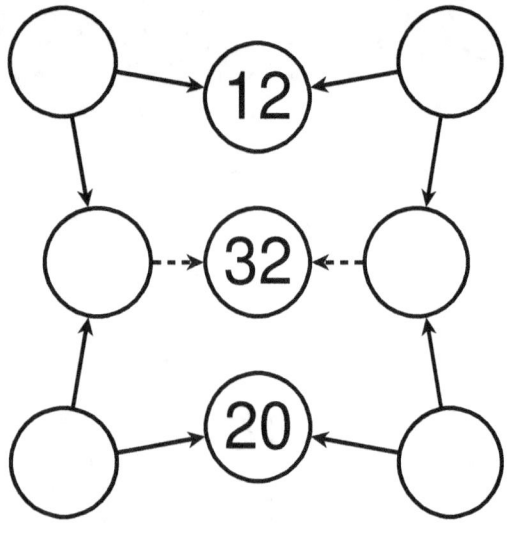

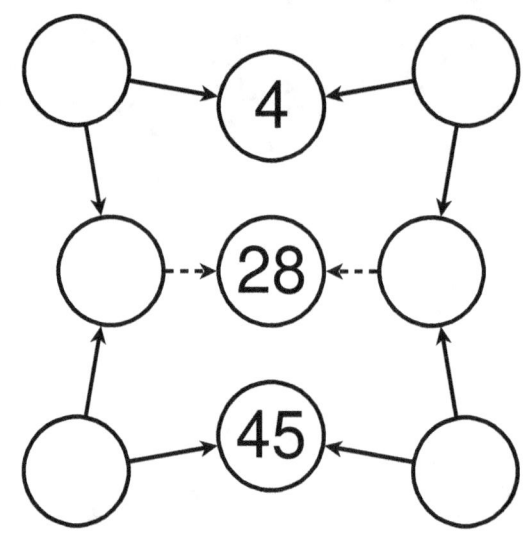

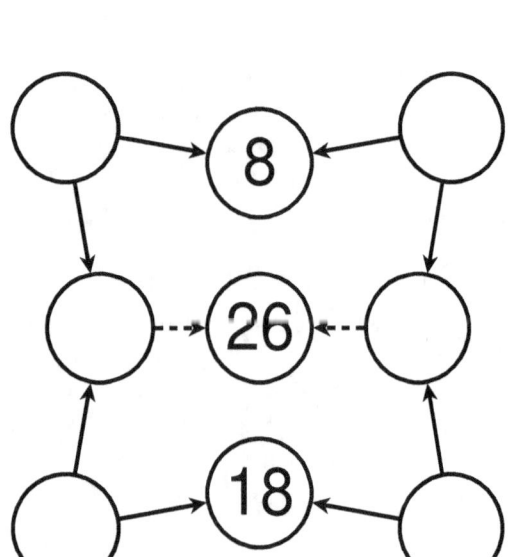

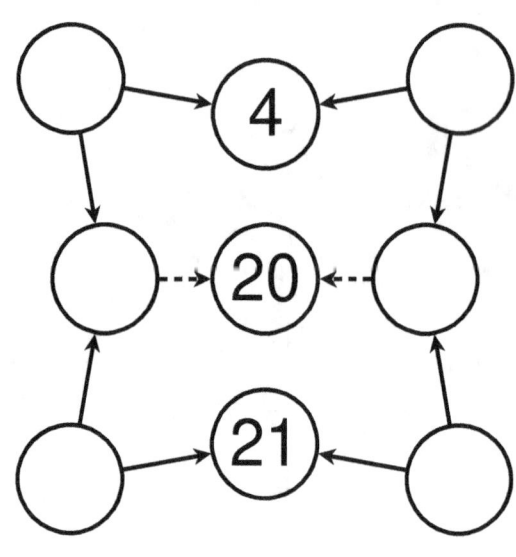

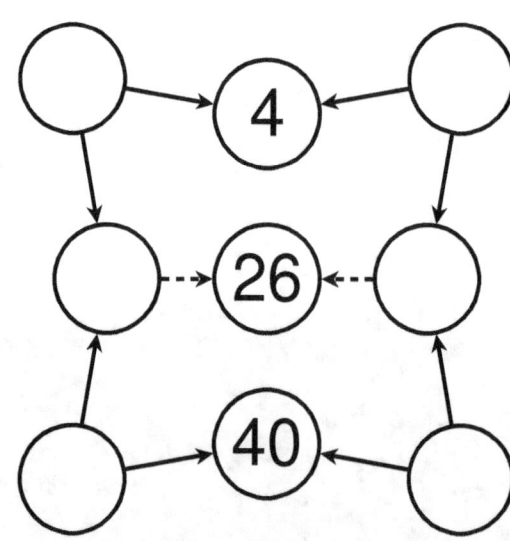

Page 58

Name _____ Date _____

Find the missing numbers. Solid lines mean multiply.
Dotted lines mean add.

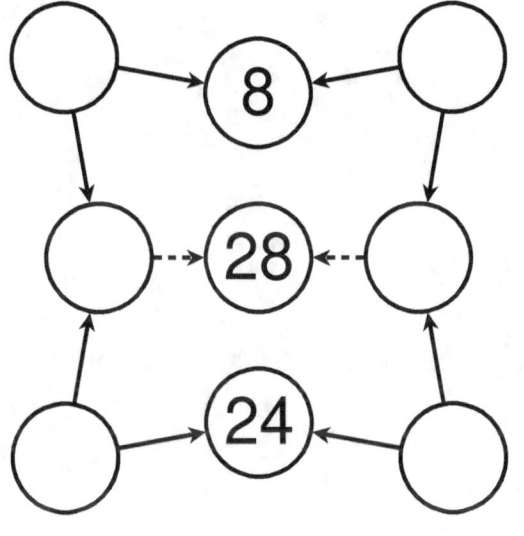

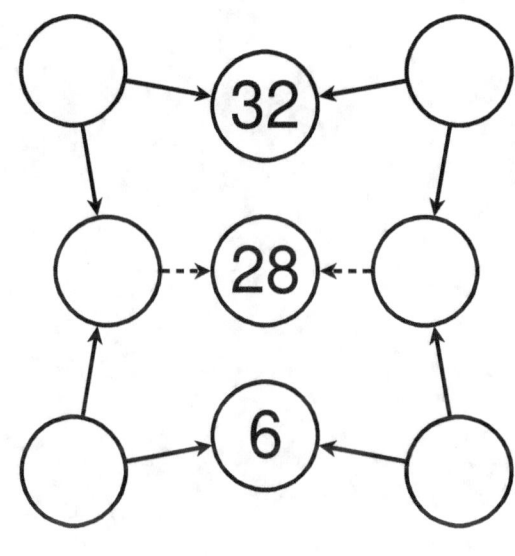

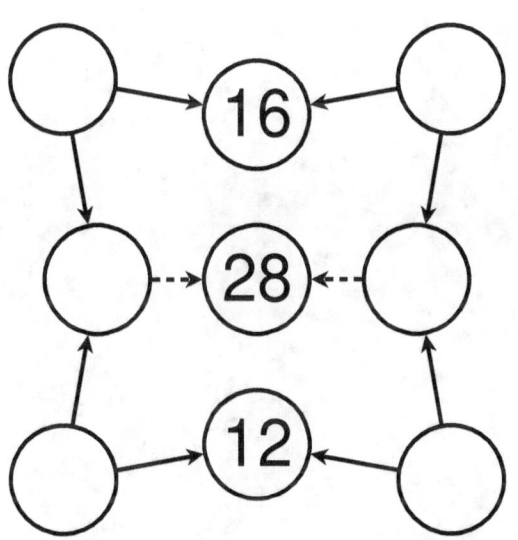

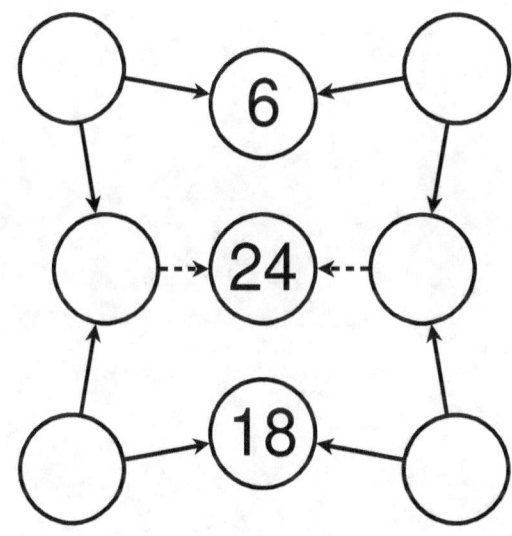

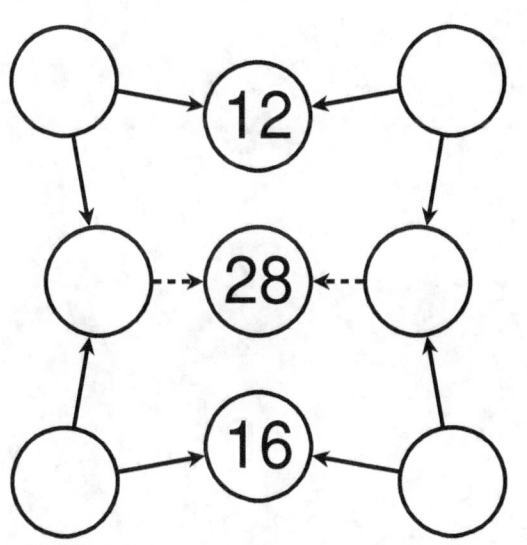

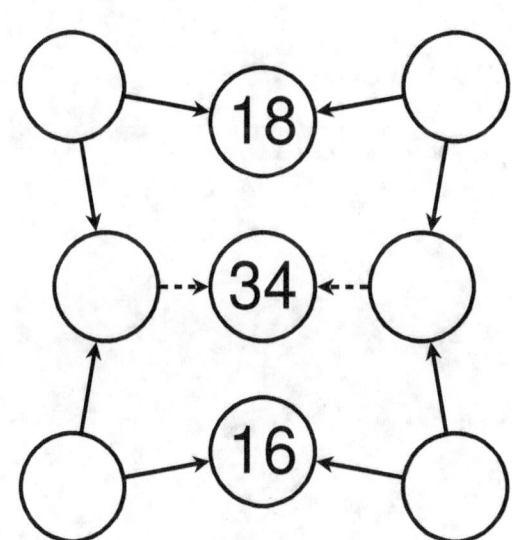

Name _____   Date _____

Find the missing numbers. Solid lines mean multiply.
Dotted lines mean add.

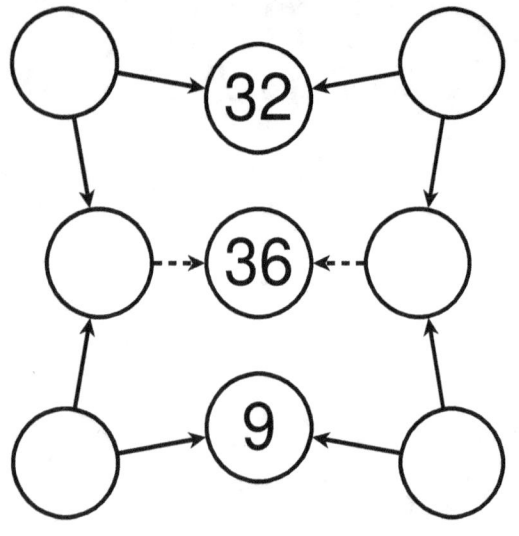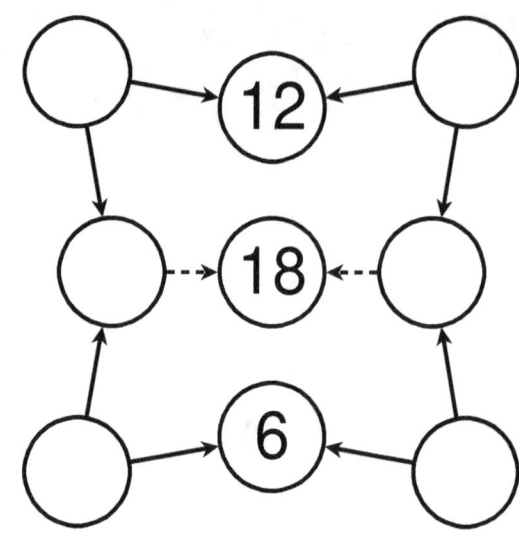

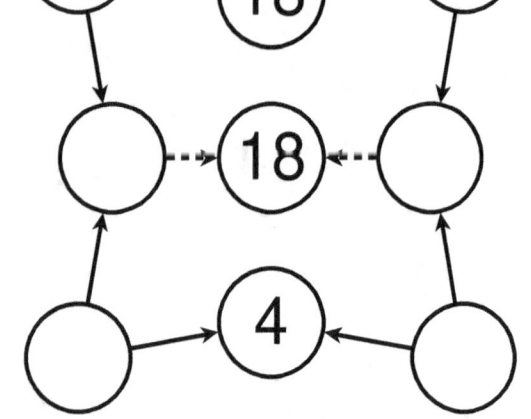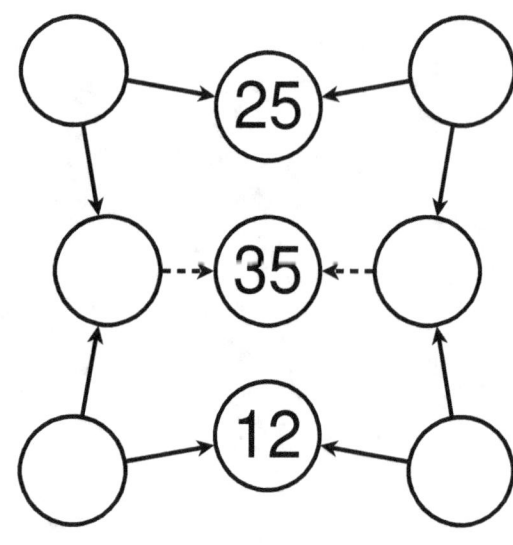

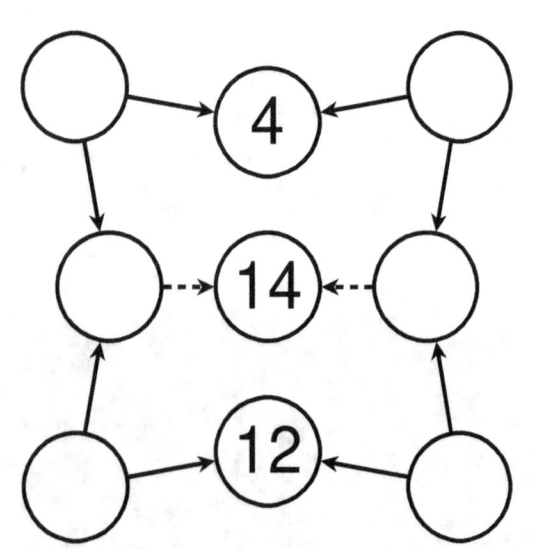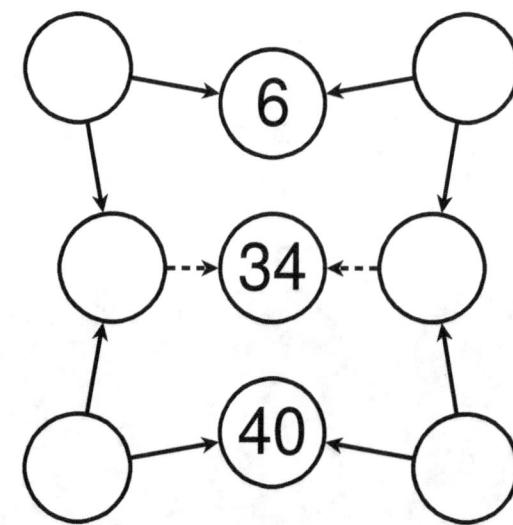

Name _____ Date _____

Find the missing numbers. Solid lines mean multiply.
Dotted lines mean add.

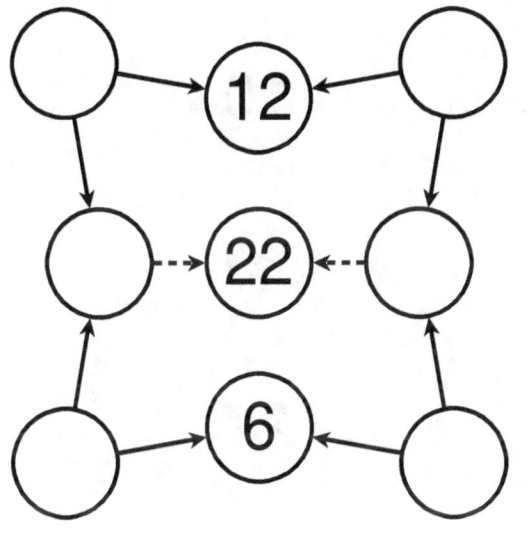

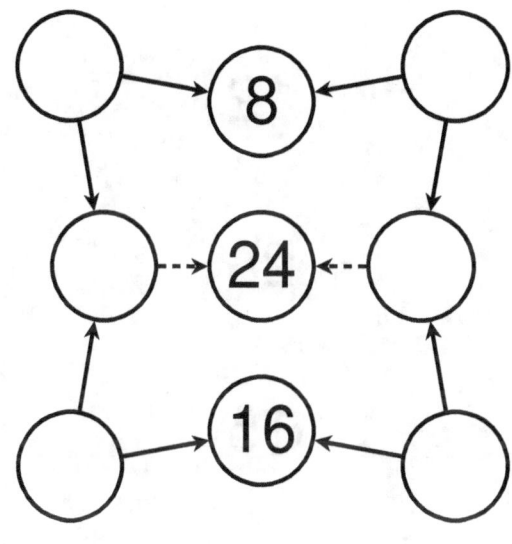

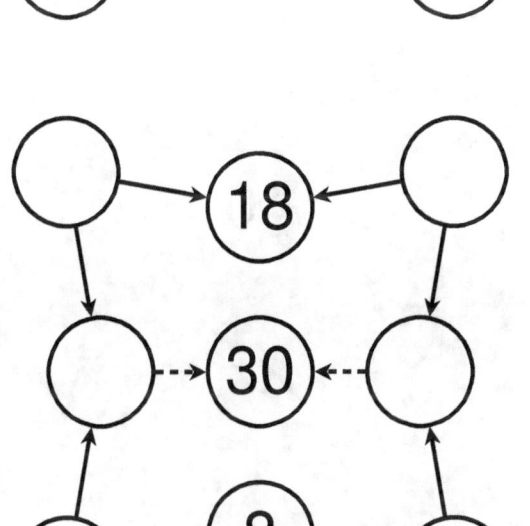

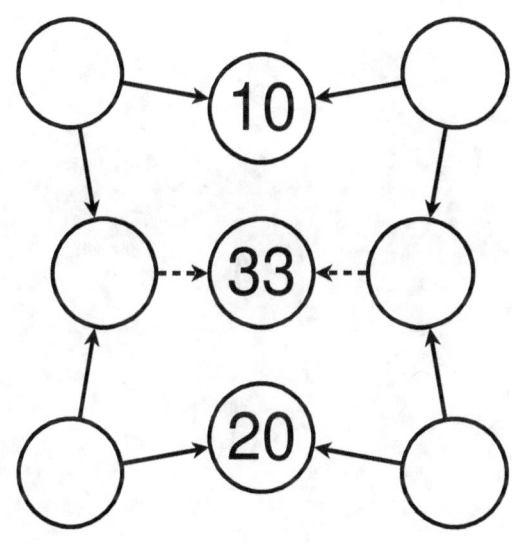

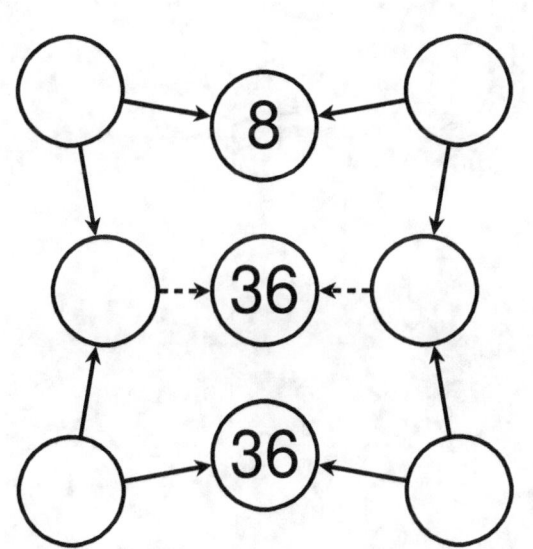

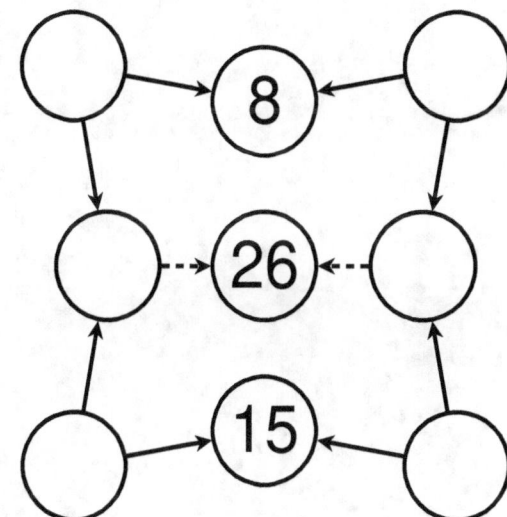

Name _____ Date _____

Find the missing numbers. Solid lines mean multiply.
Dotted lines mean add.

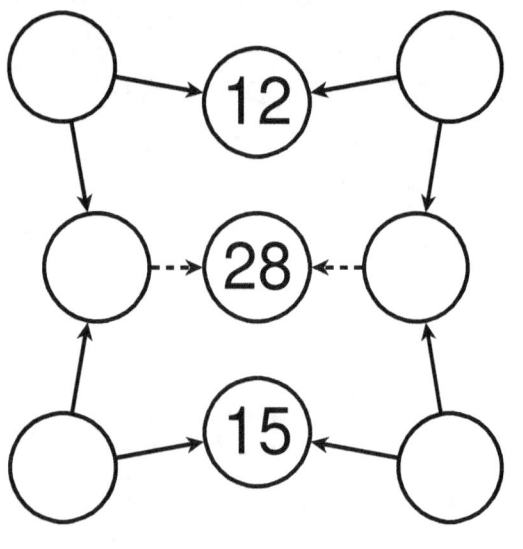

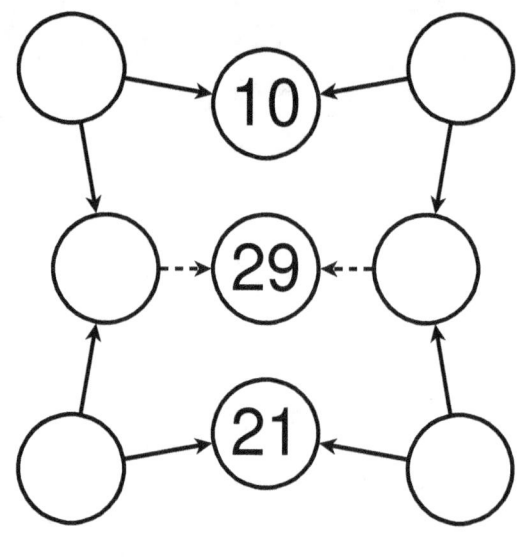

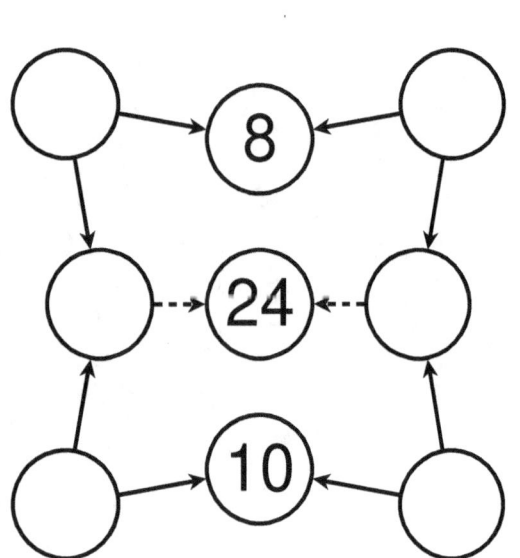

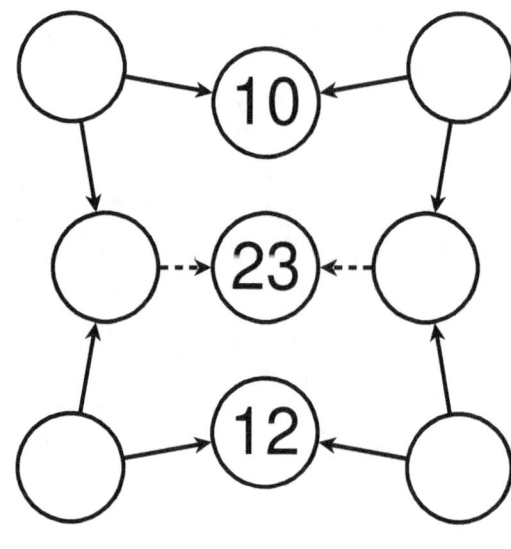

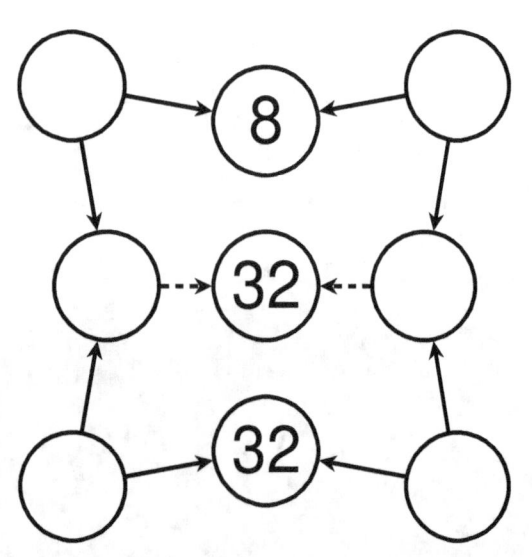

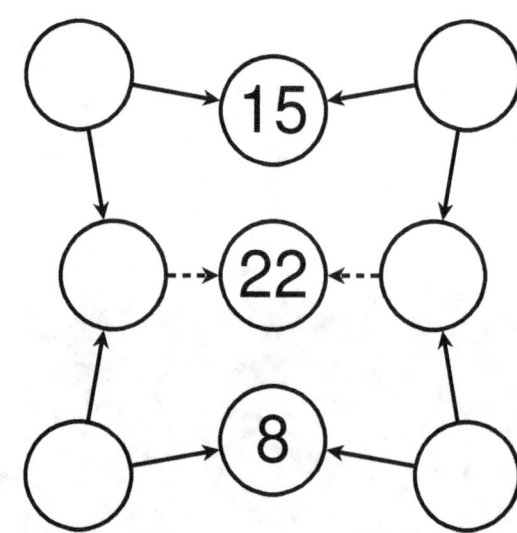

Page 62

Name _____ Date _____

Find the missing numbers. Solid lines mean multiply.
Dotted lines mean add.

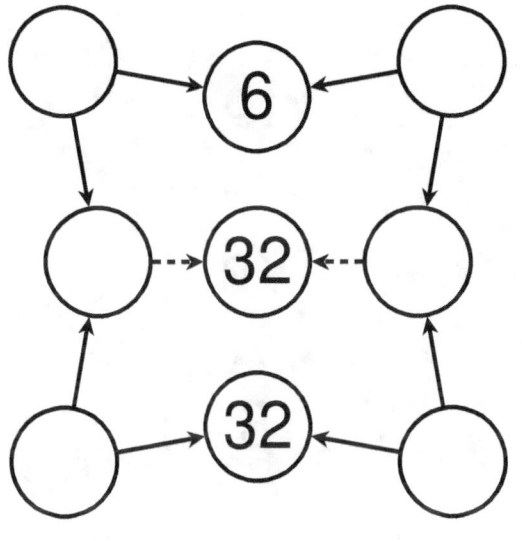

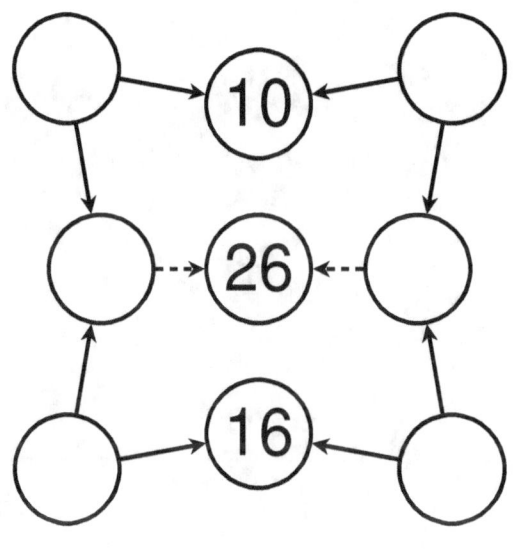

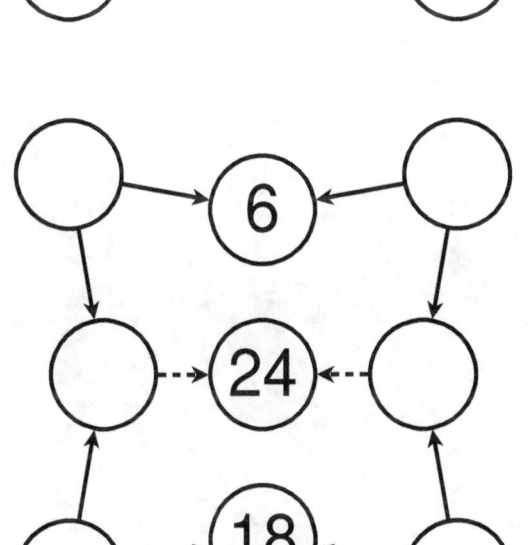

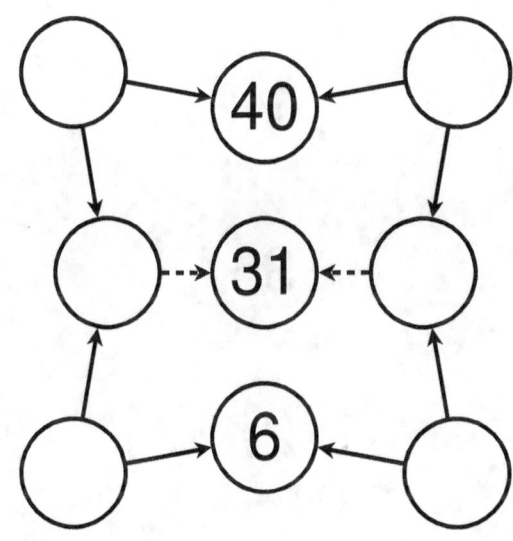

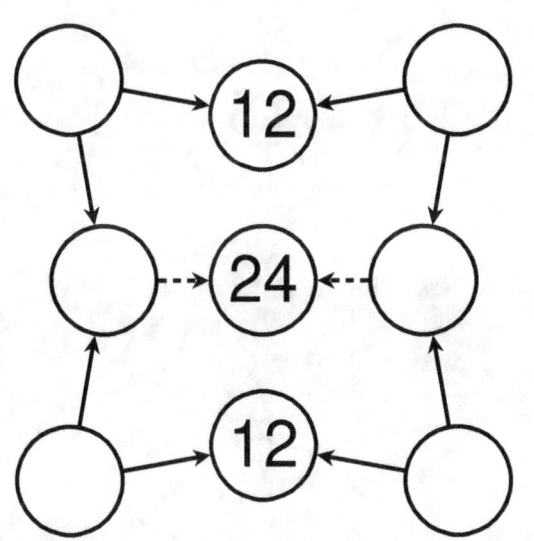

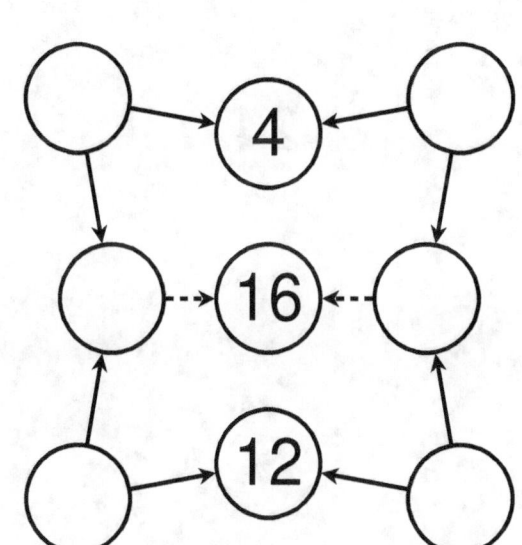

Name _____ Date _____

Find the missing numbers. Solid lines mean multiply.
Dotted lines mean add.

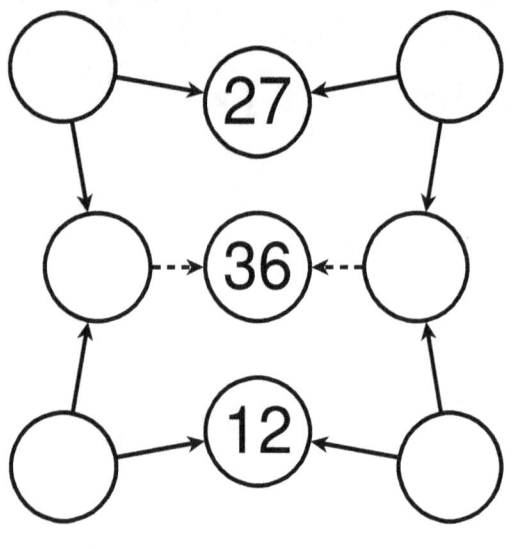

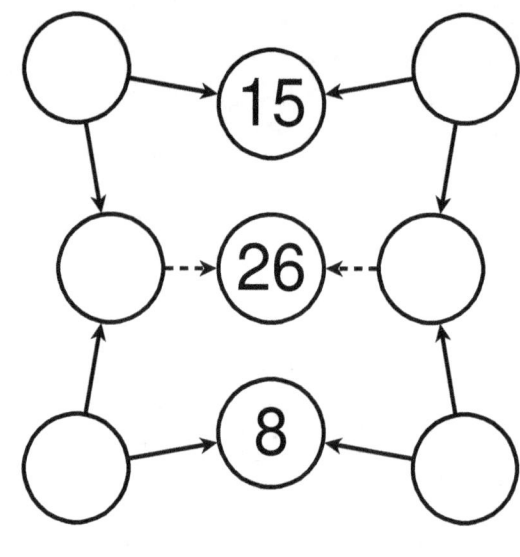

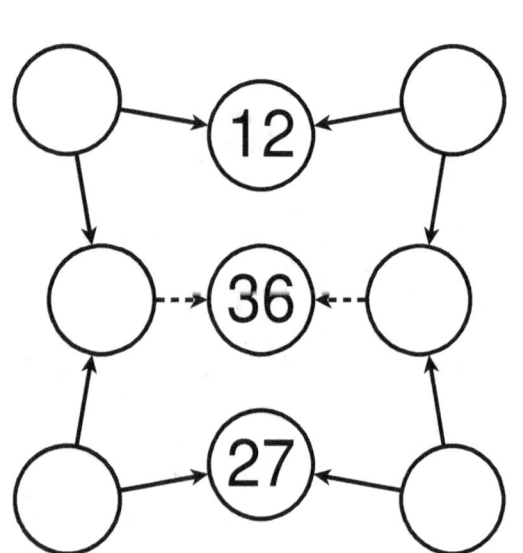

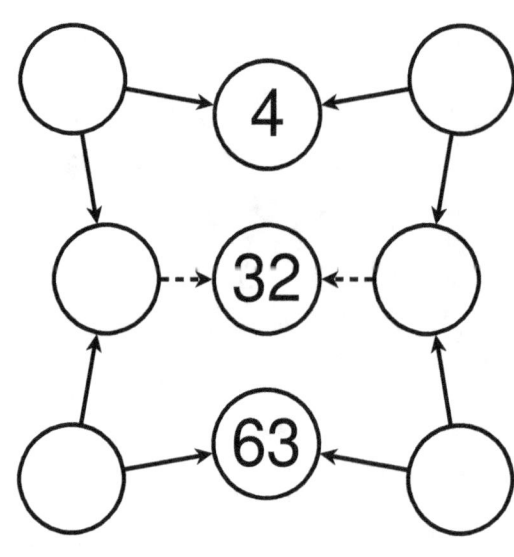

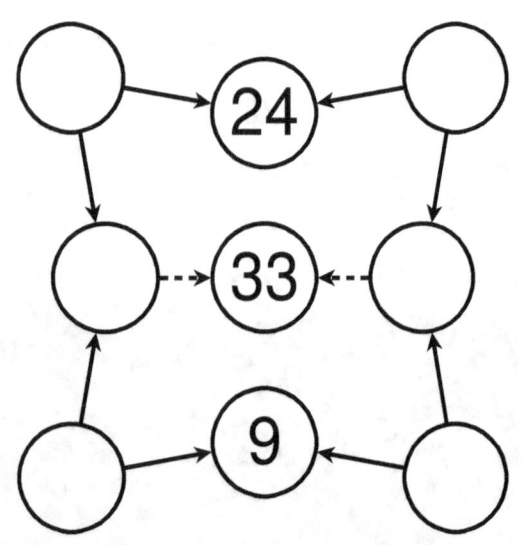

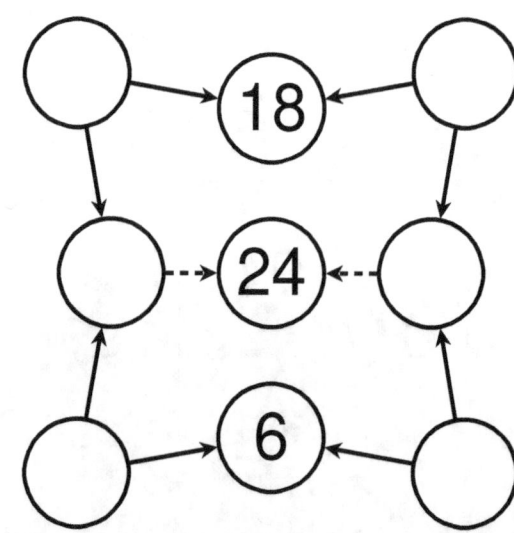

Name _____ Date _____

Find the missing numbers. Solid lines mean multiply.
Dotted lines mean add.

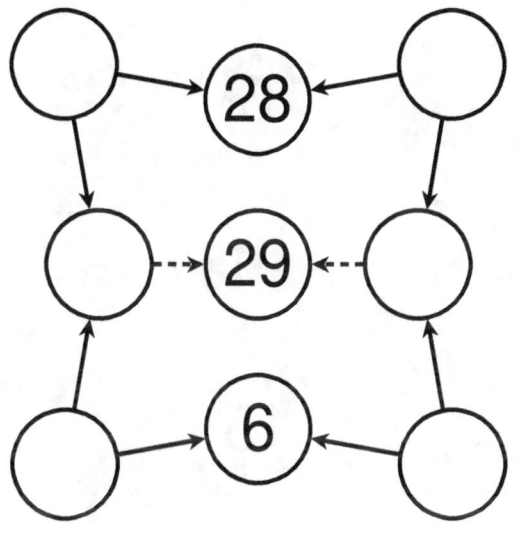

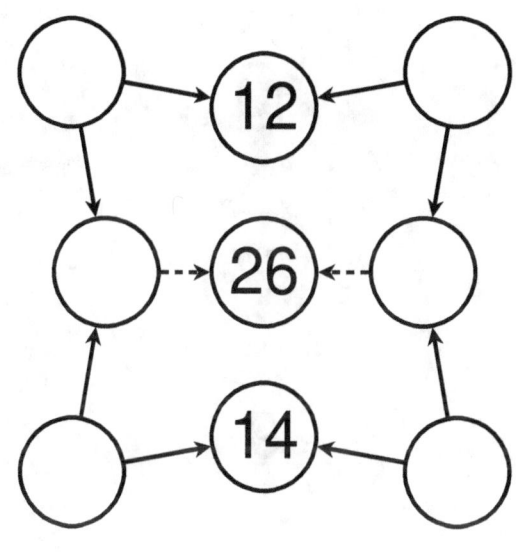

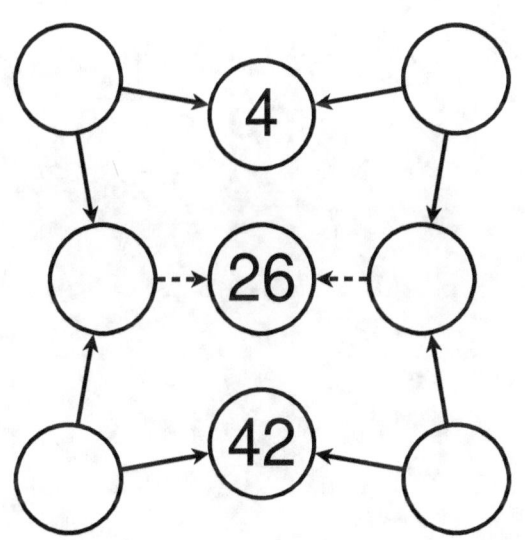

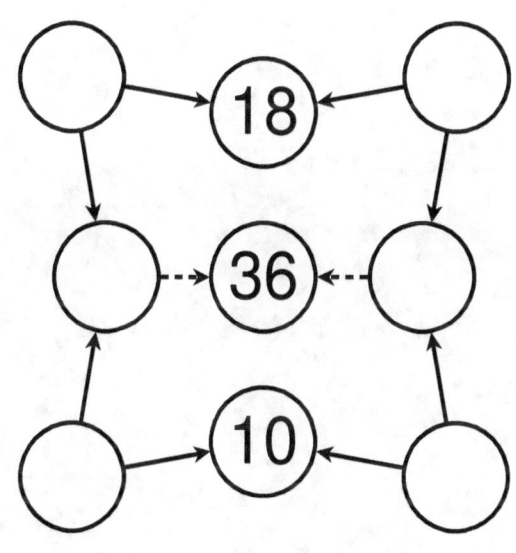

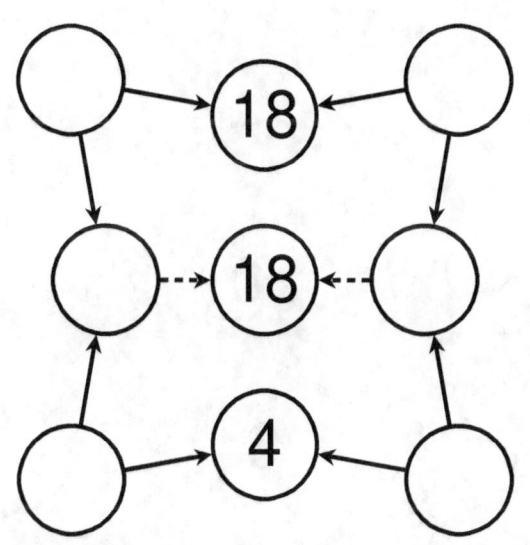

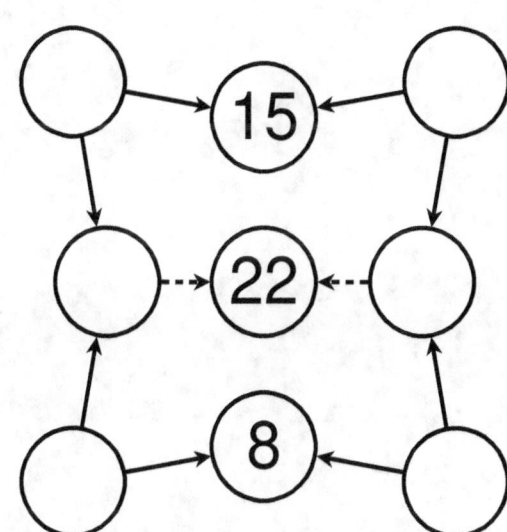

Name _____ Date _____

Find the missing numbers. Solid lines mean multiply.
Dotted lines mean add.

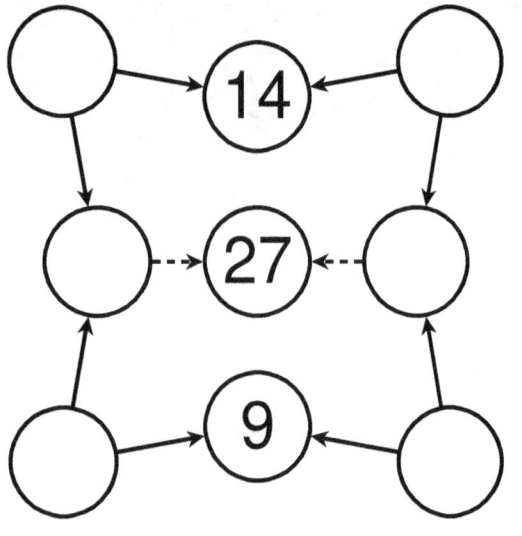

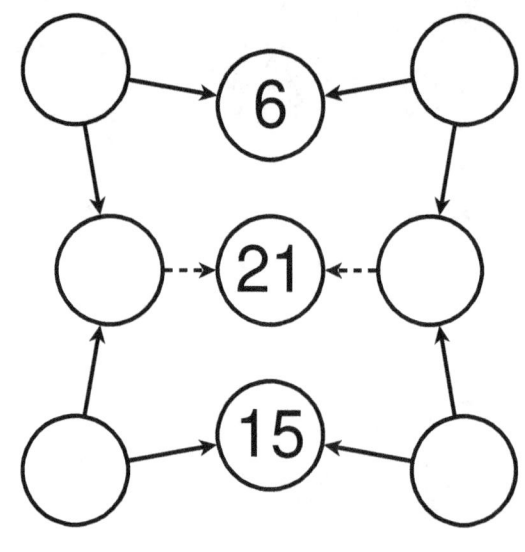

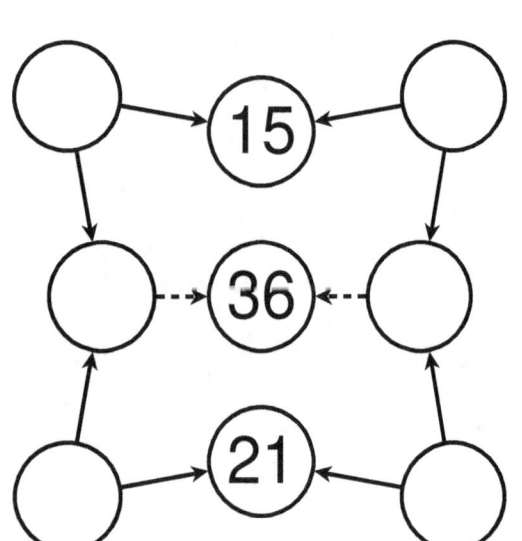

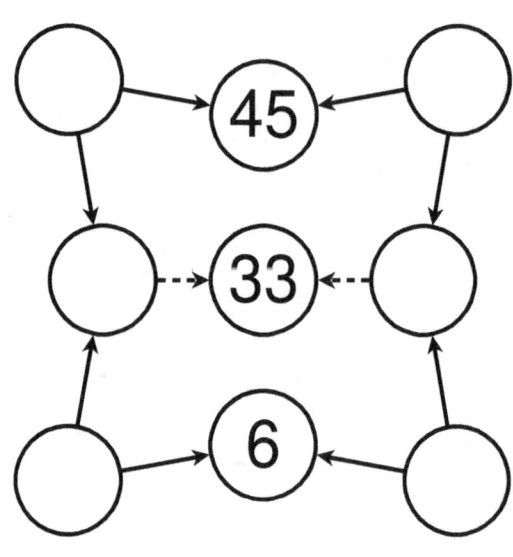

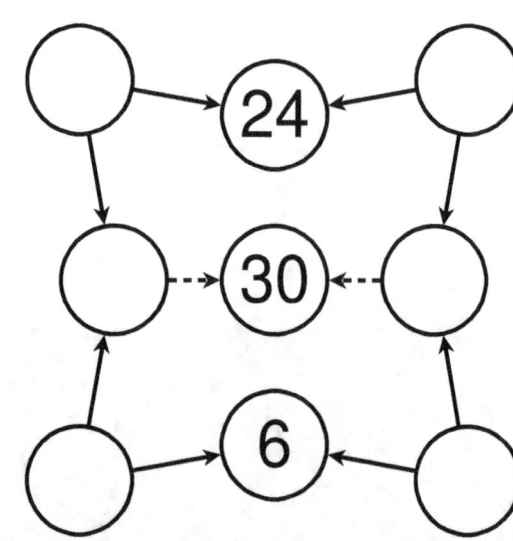

Name _____ Date _____

Find the missing numbers. Solid lines mean multiply.
Dotted lines mean add.

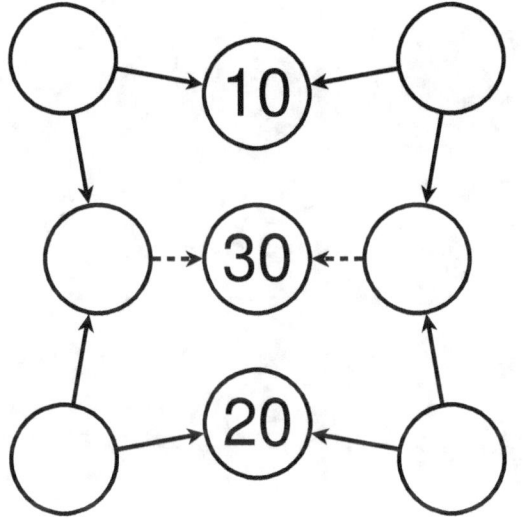

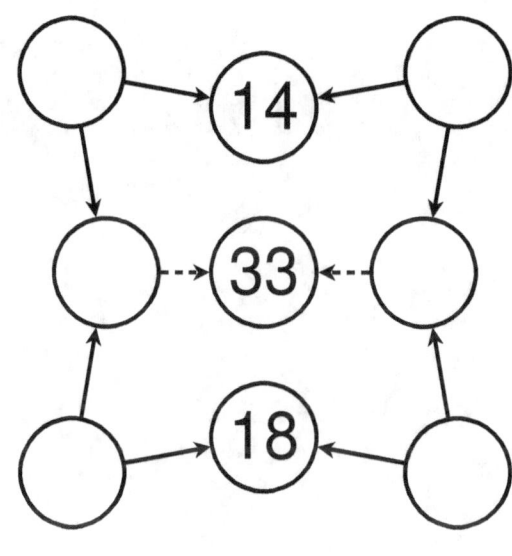

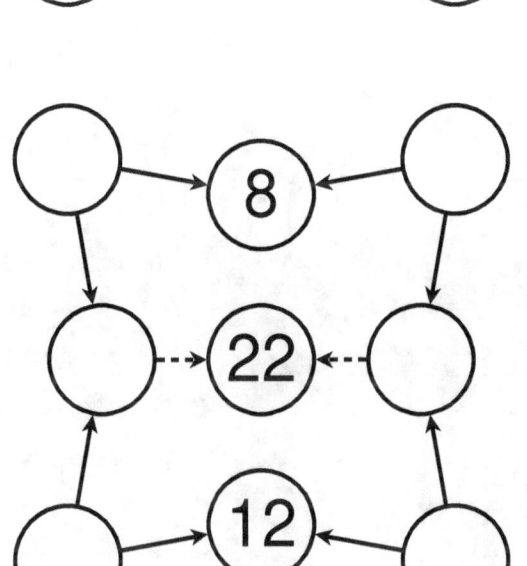

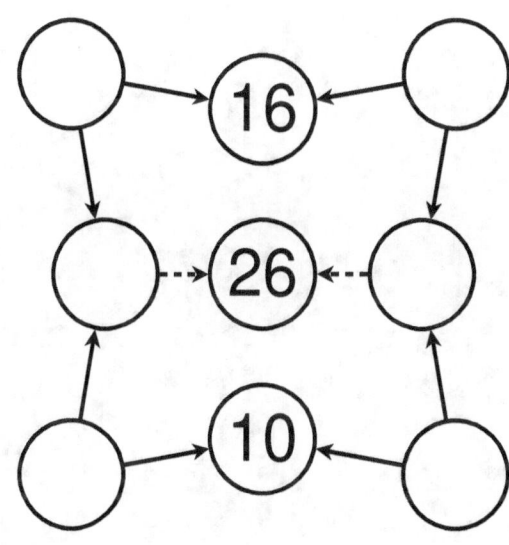

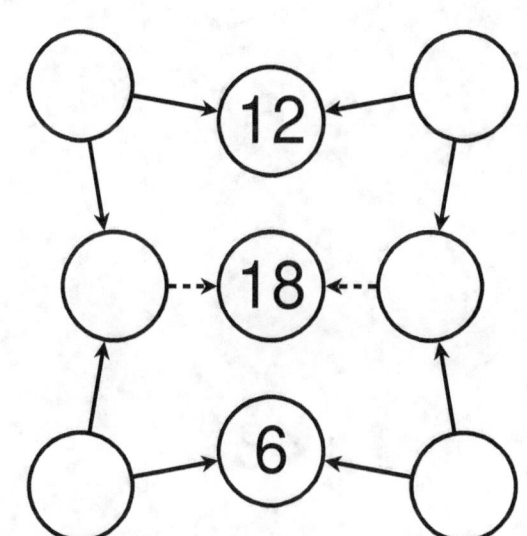

Name _____ Date _____

Find the missing numbers. Solid lines mean multiply.
Dotted lines mean add.

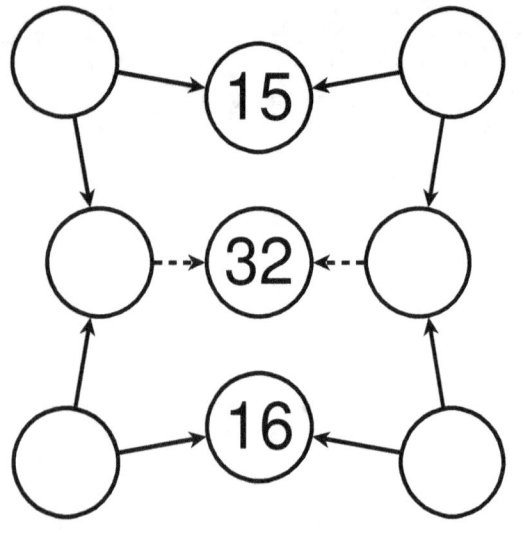

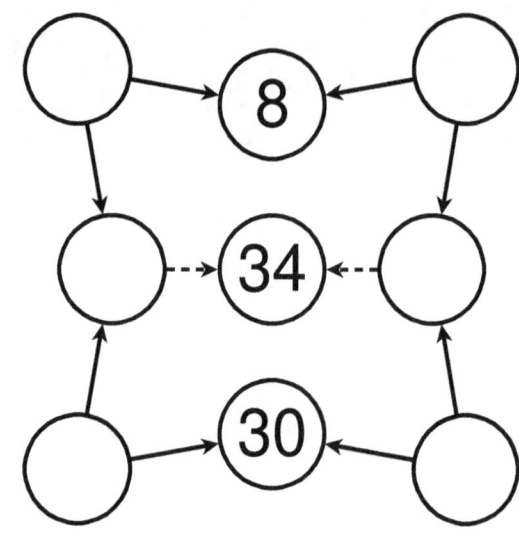

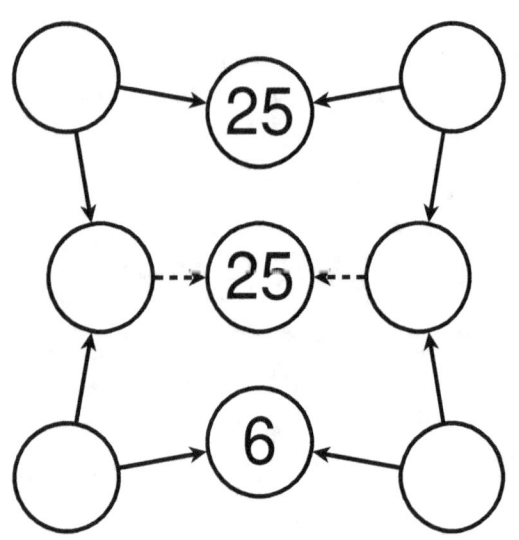

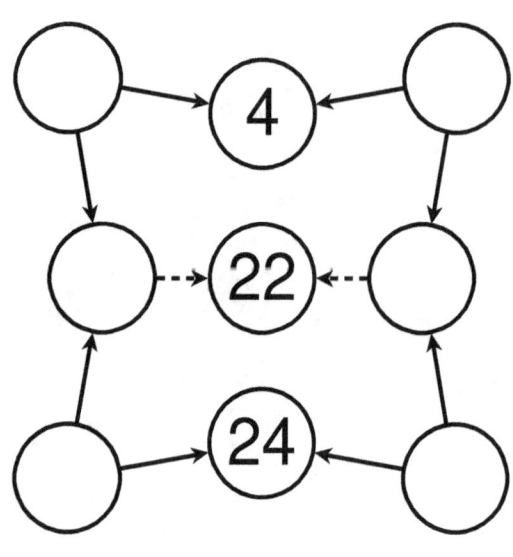

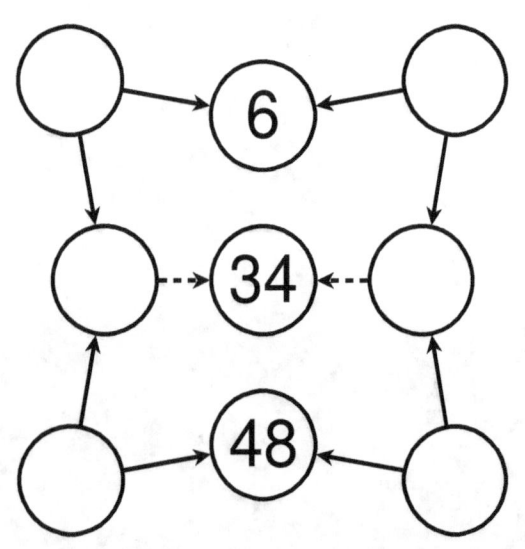

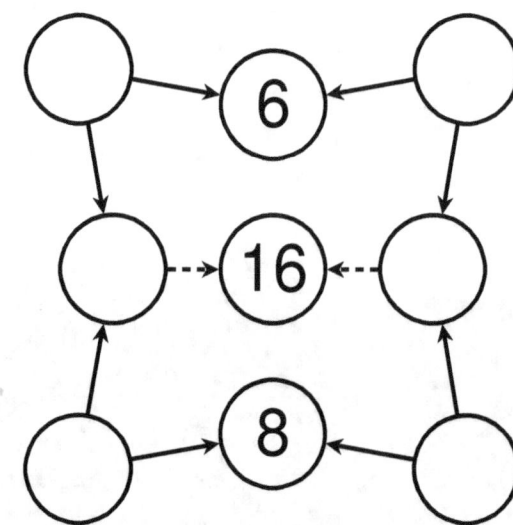

Name _____ Date _____

Find the missing numbers. Solid lines mean multiply.
Dotted lines mean add.

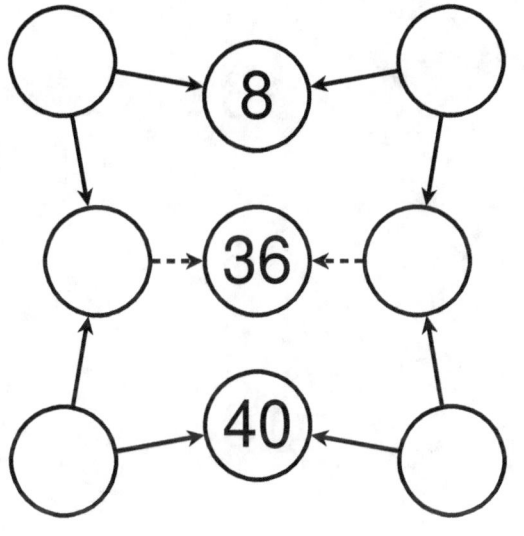

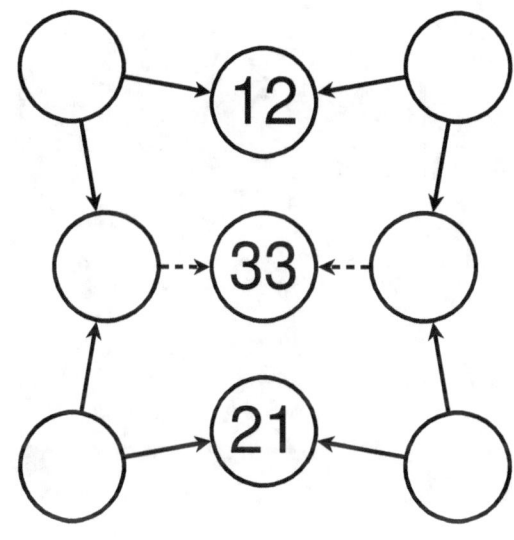

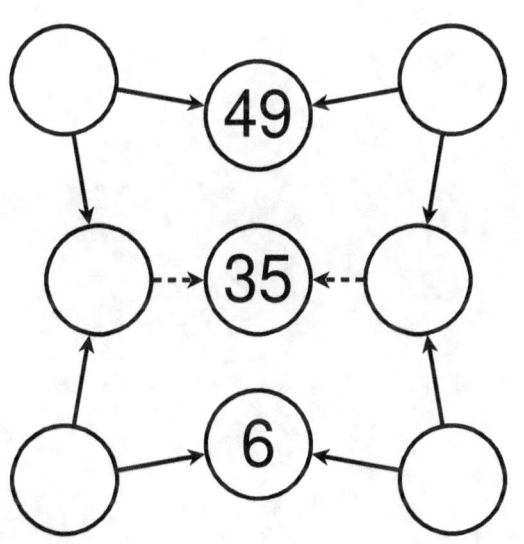

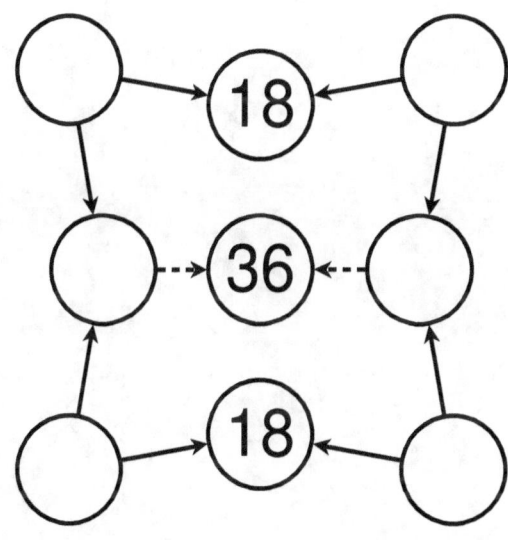

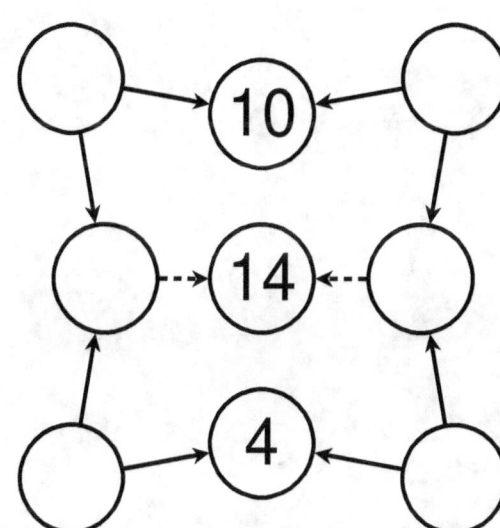

Name _____   Date _____

Find the missing numbers. Solid lines mean multiply.
Dotted lines mean add.

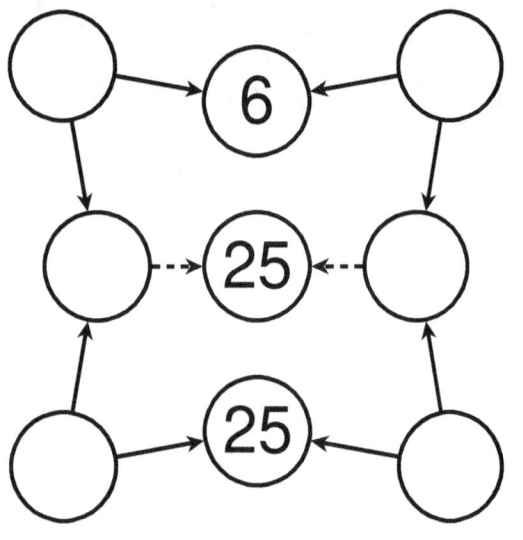

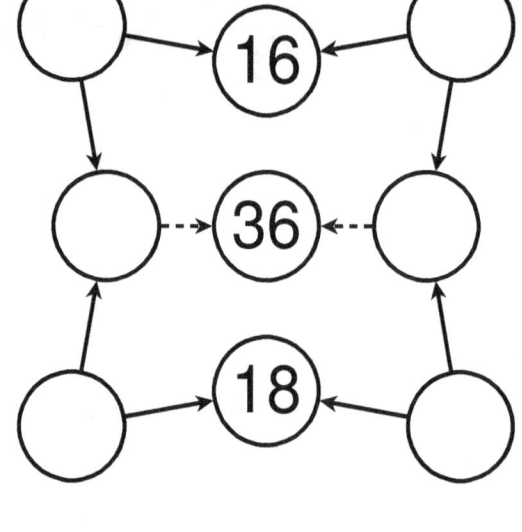

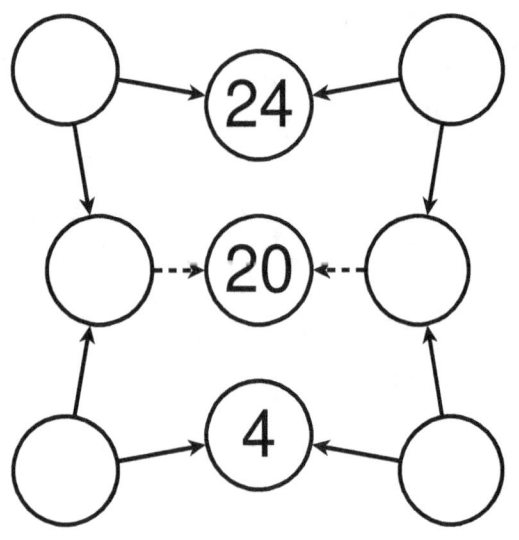

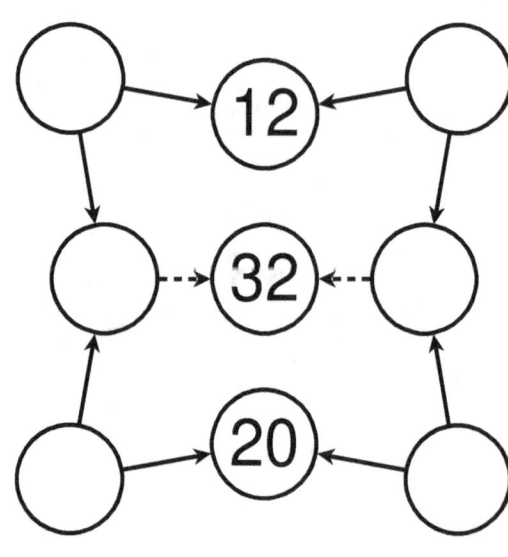

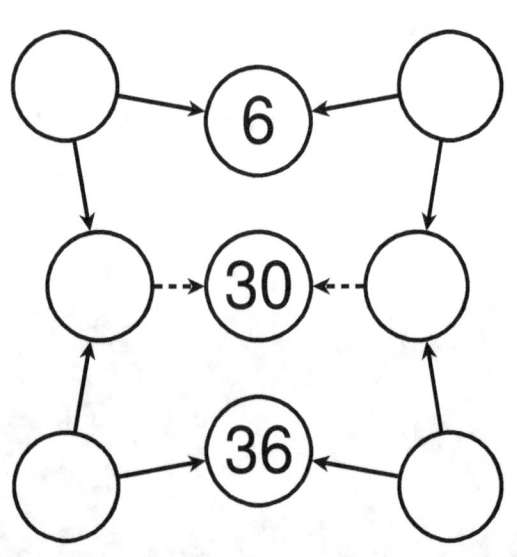

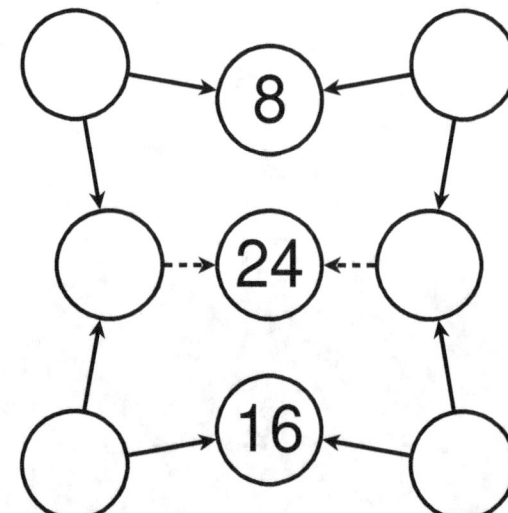

Find the missing numbers. Solid lines mean multiply.
Dotted lines mean add.

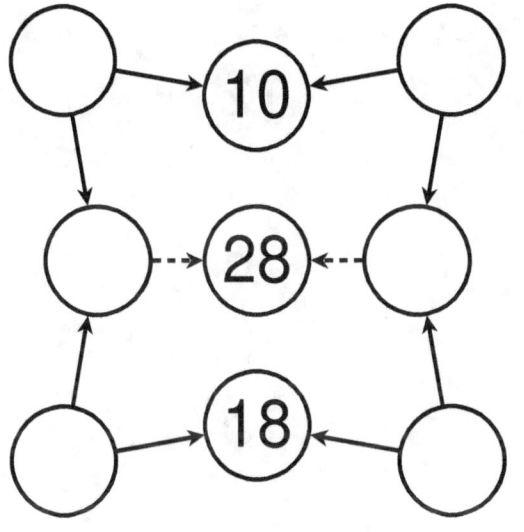

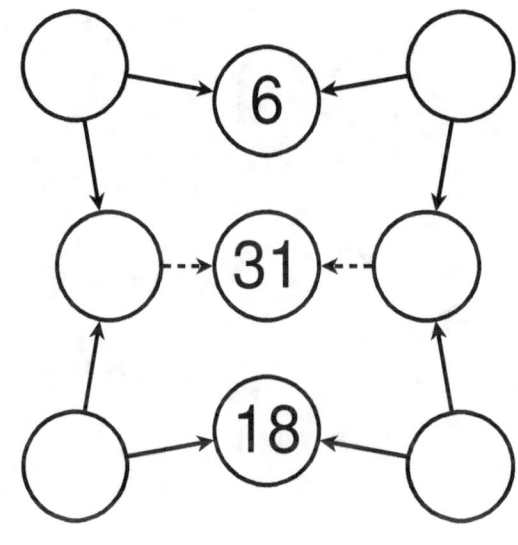

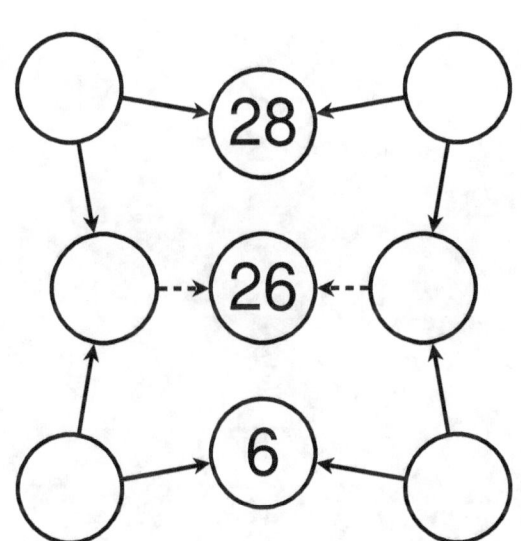

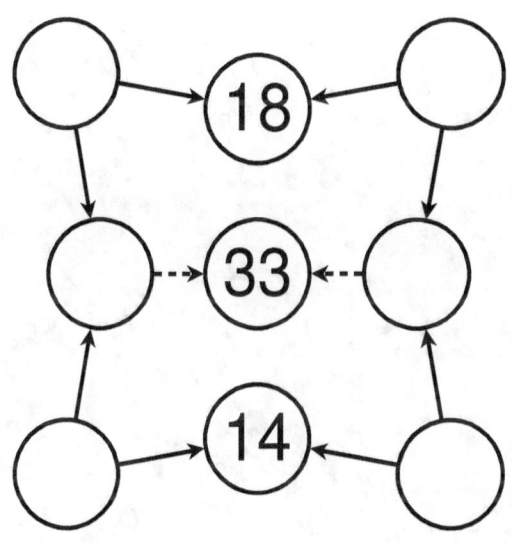

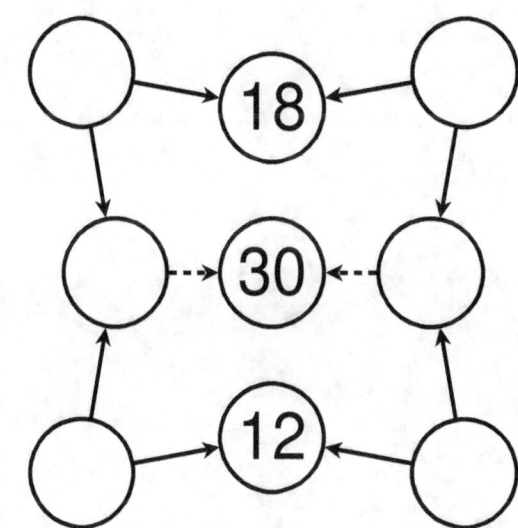

Name _____ Date _____

Find the missing numbers. Solid lines mean multiply.
Dotted lines mean add.

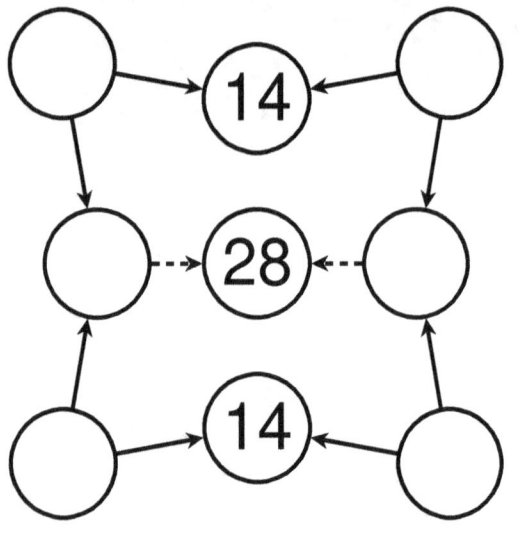

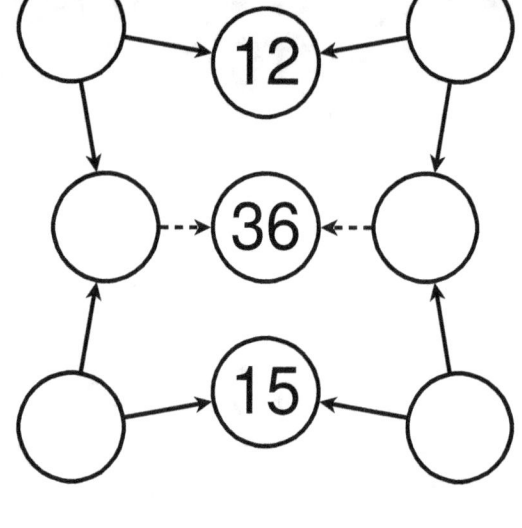

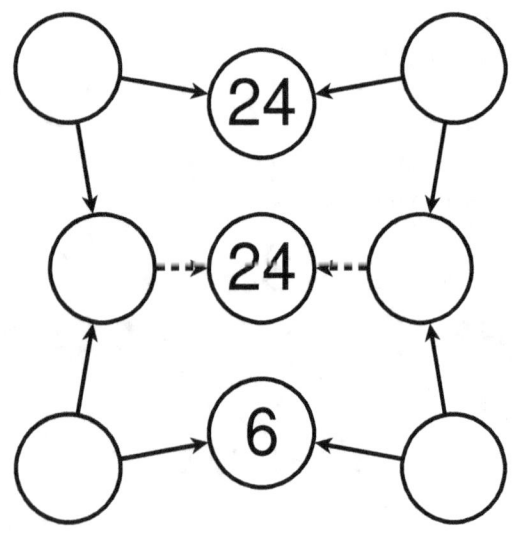

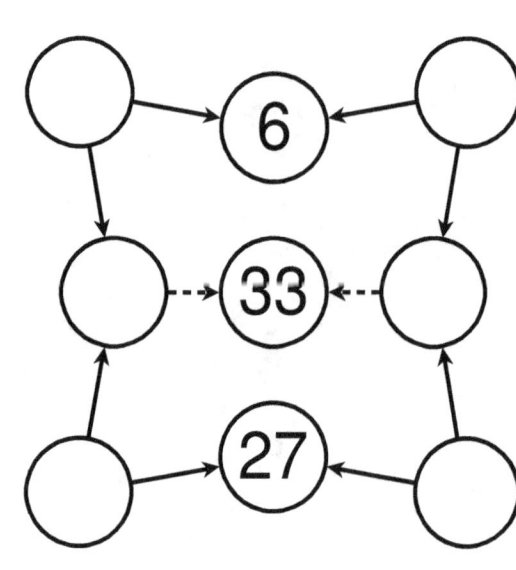

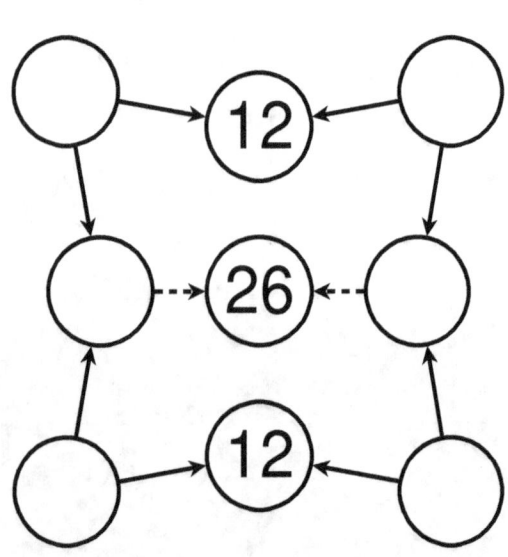

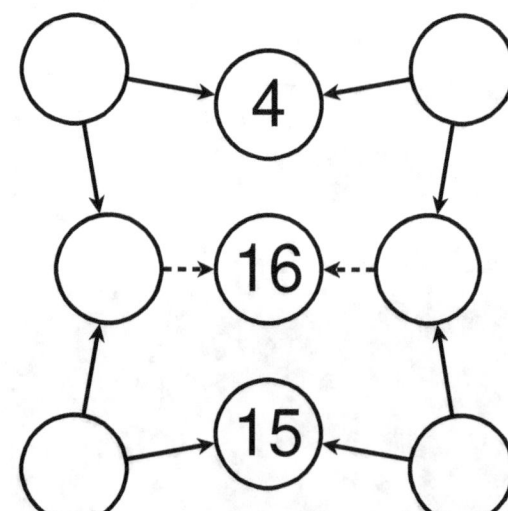

Name _____ Date _____

Find the missing numbers. Solid lines mean multiply.
Dotted lines mean add.

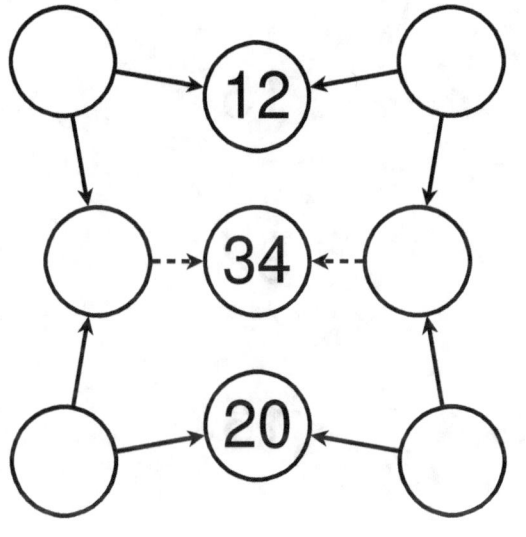

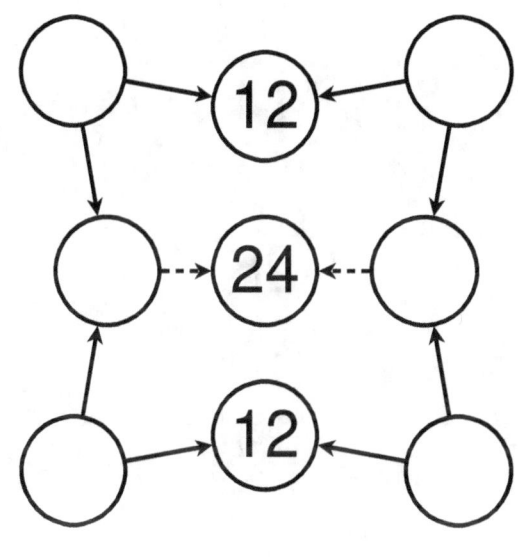

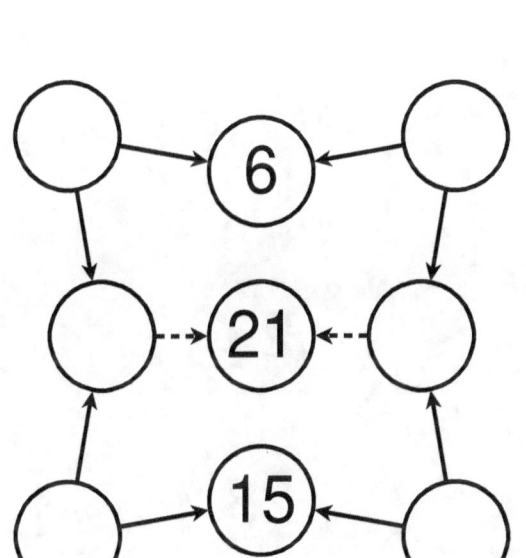

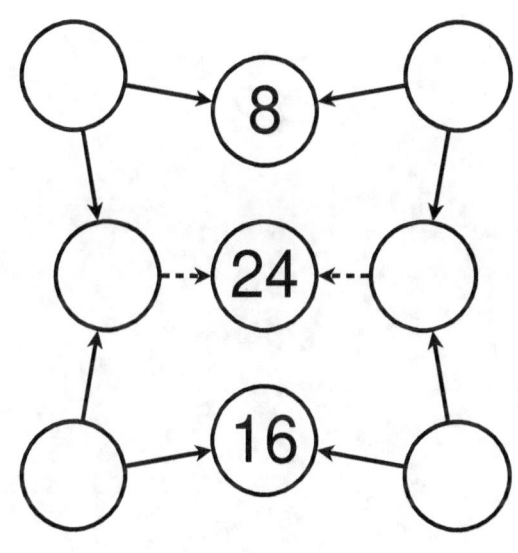

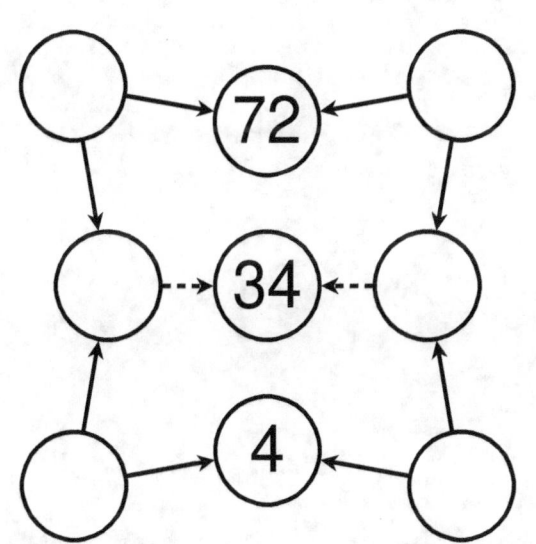

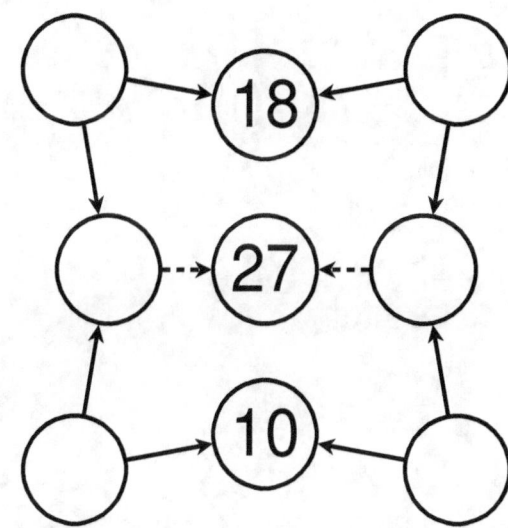

Name _____ Date _____

Find the missing numbers. Solid lines mean multiply.
Dotted lines mean add.

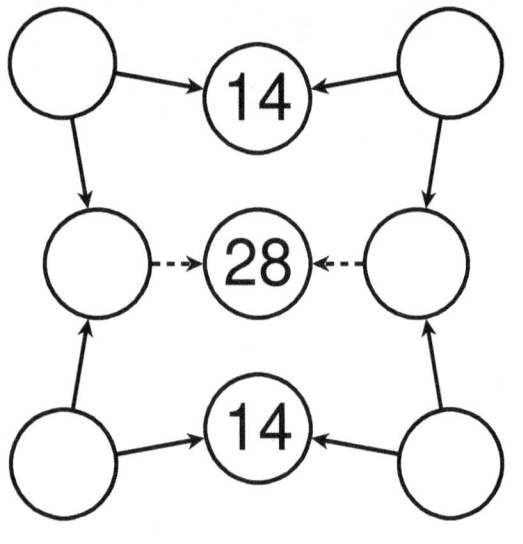

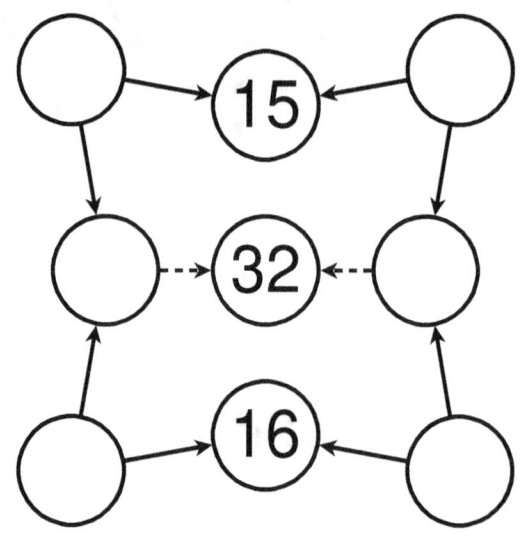

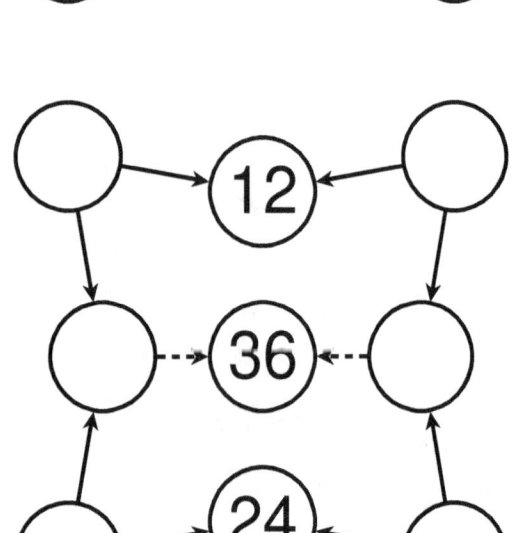

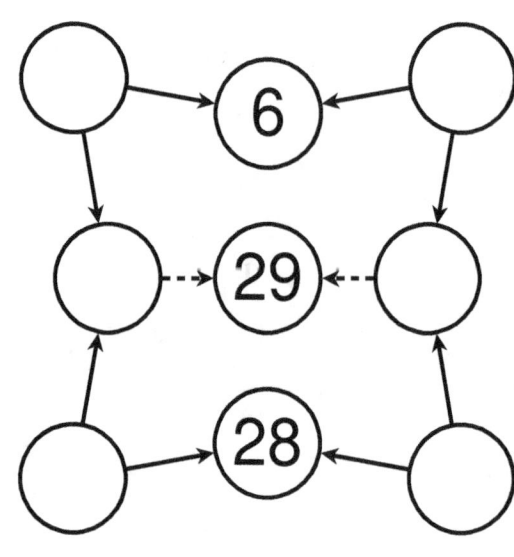

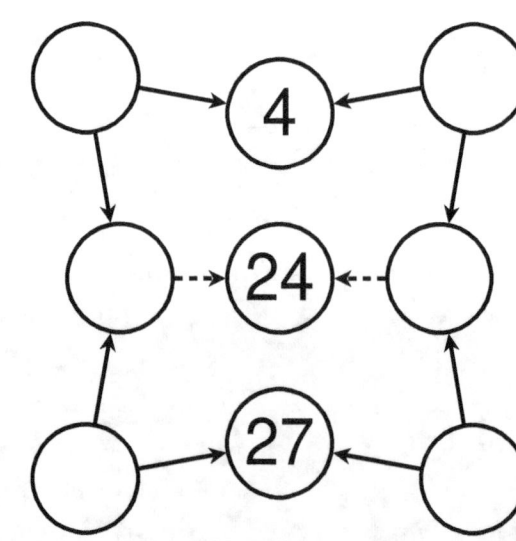

Name _____ Date _____

Find the missing numbers. Solid lines mean multiply.
Dotted lines mean add.

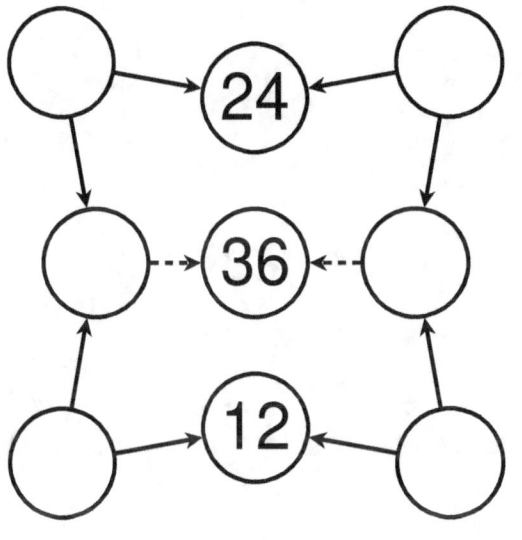

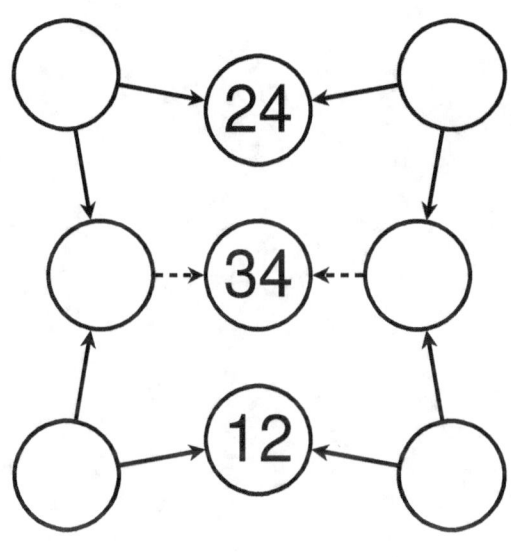

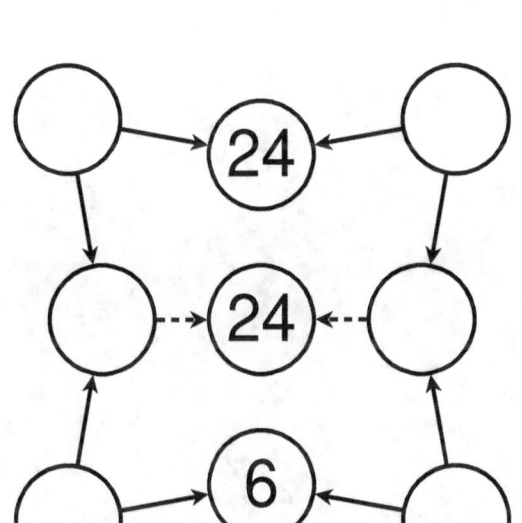

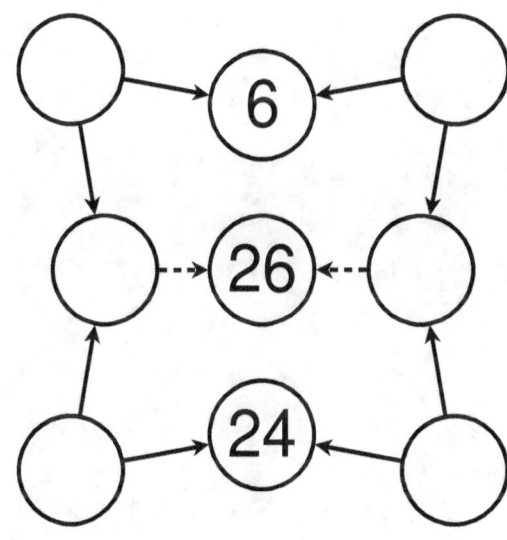

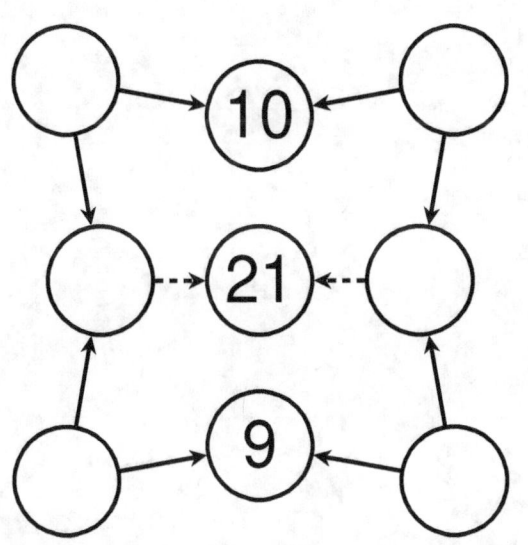

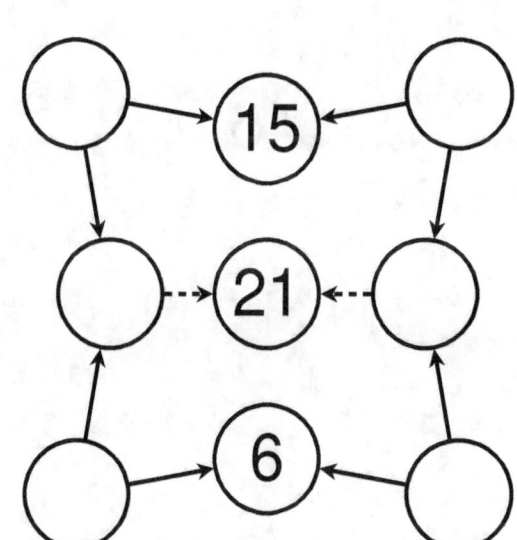

Page 75

Name _____ Date _____

Find the missing numbers. Solid lines mean multiply.
Dotted lines mean add.

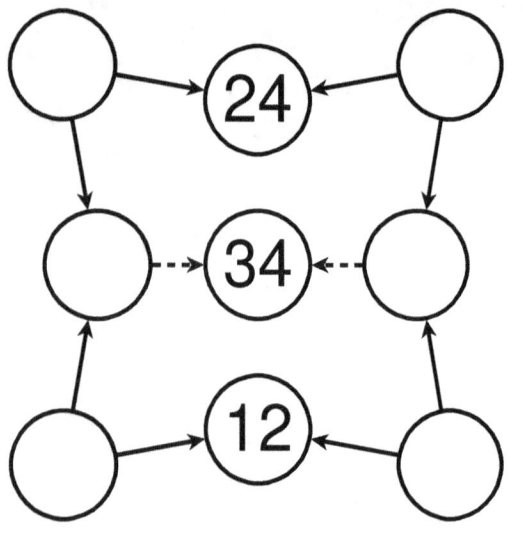

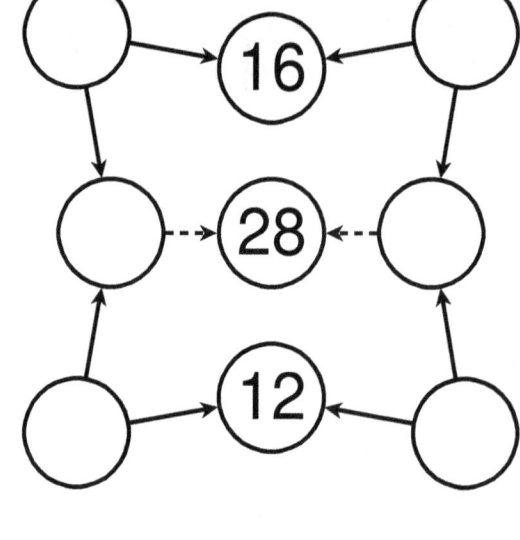

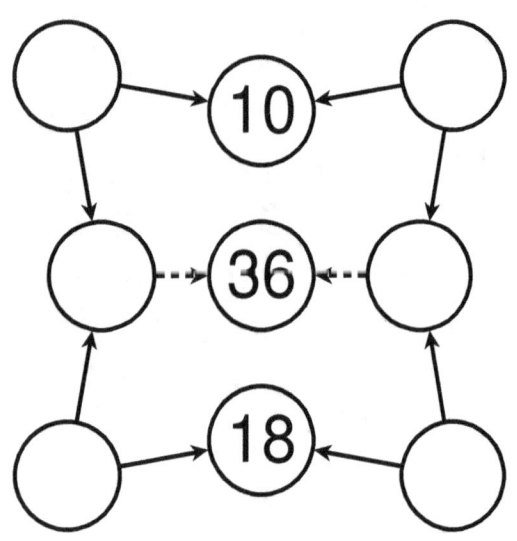

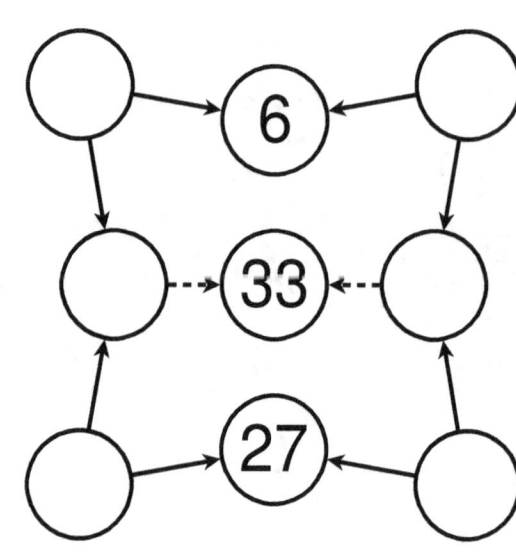

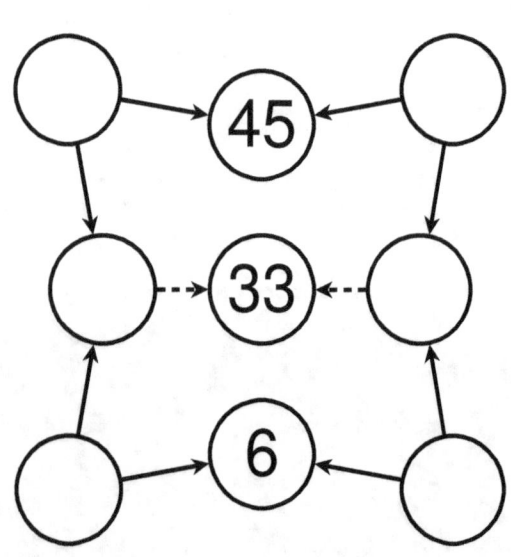

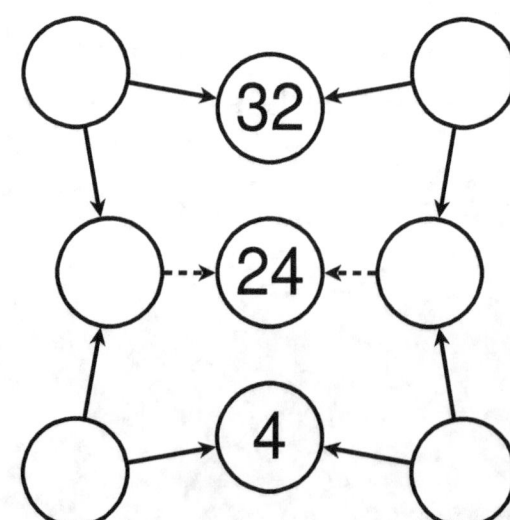

Name _____ Date _____

Find the missing numbers. Solid lines mean multiply.
Dotted lines mean add.

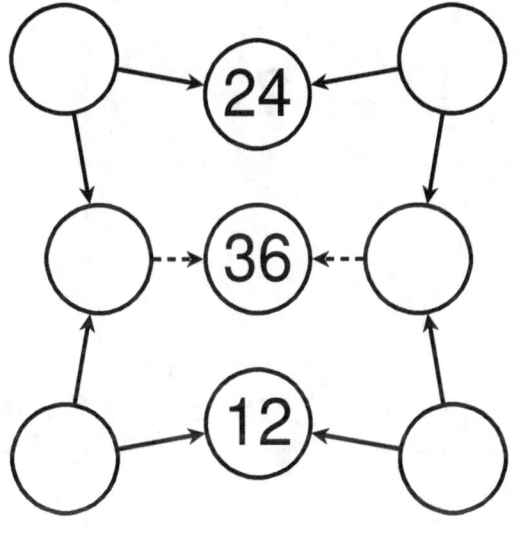

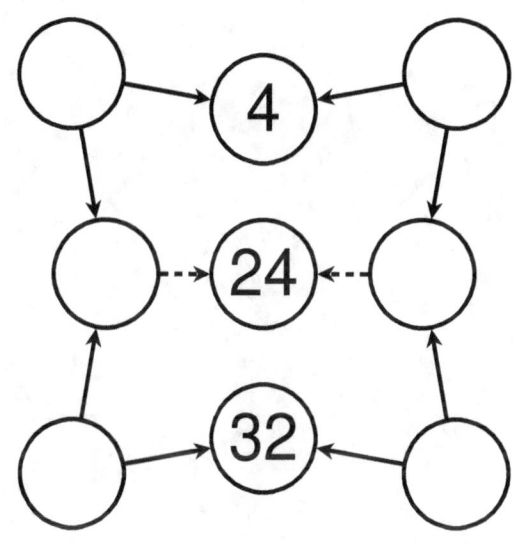

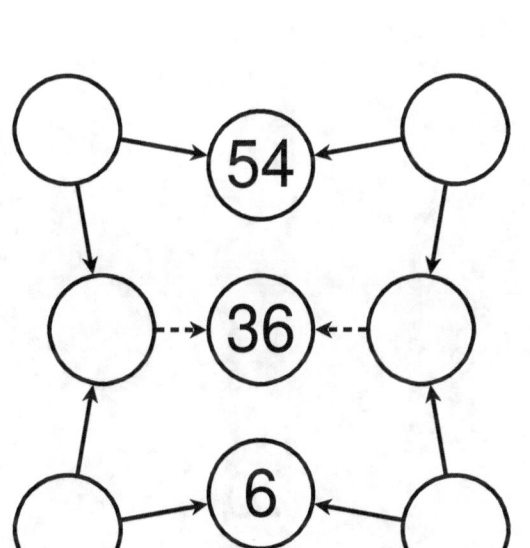

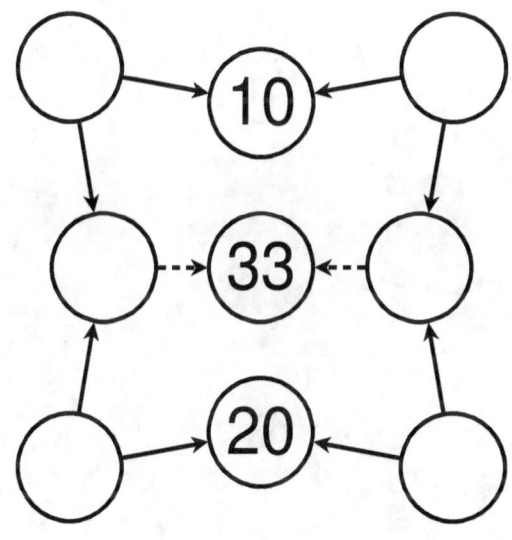

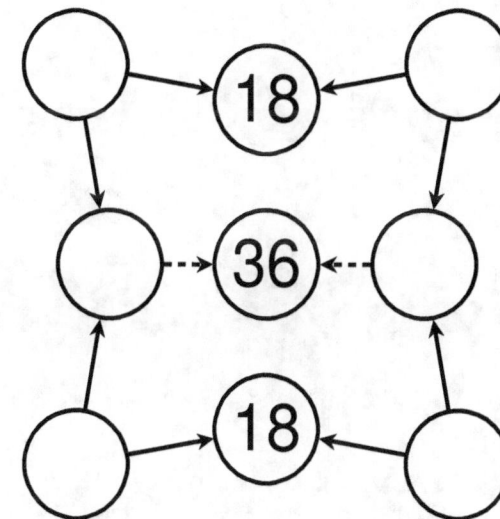

Page 77

Name _____ Date _____

Find the missing numbers. Solid lines mean multiply.
Dotted lines mean add.

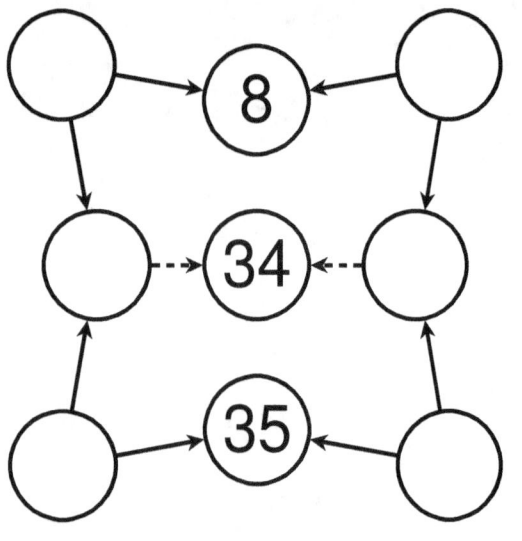

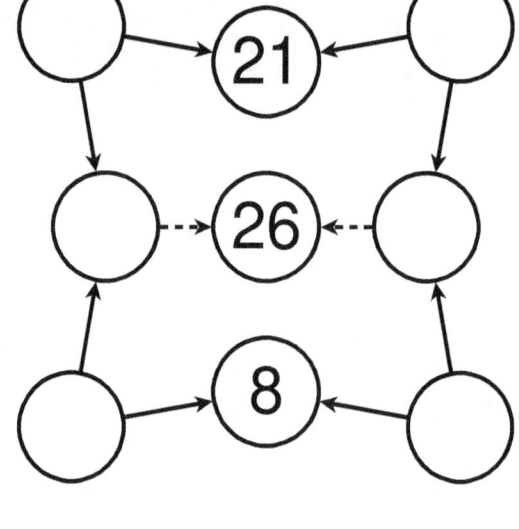

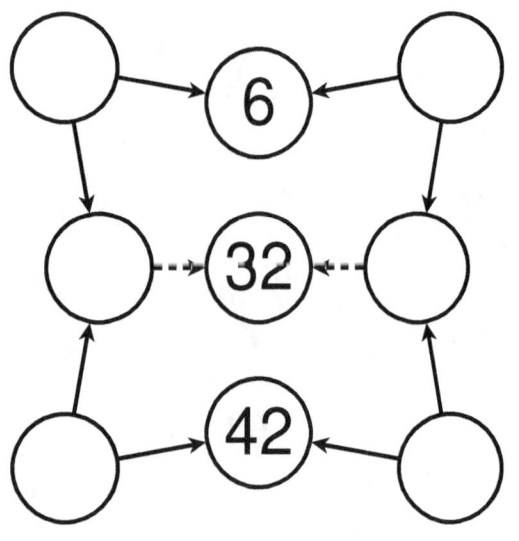

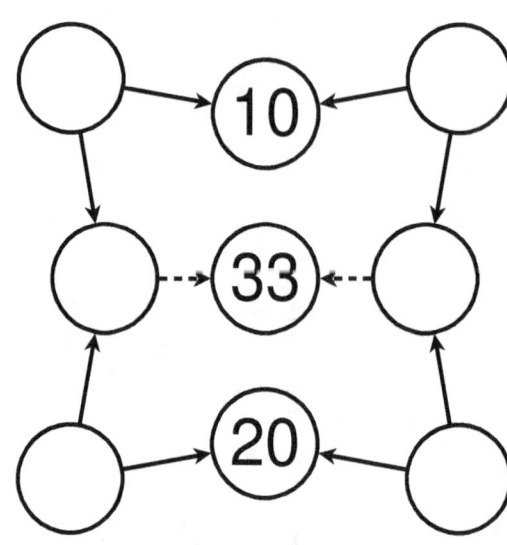

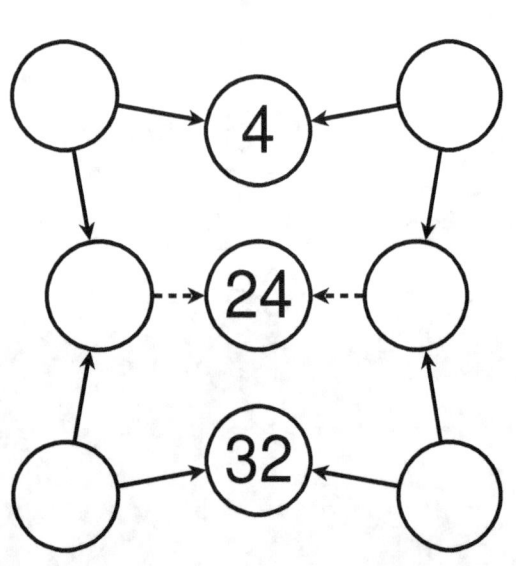

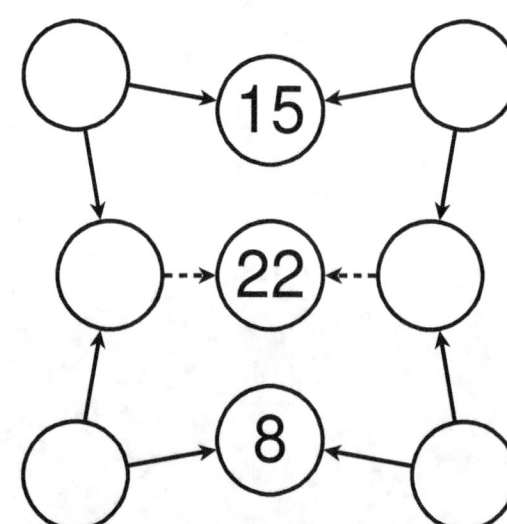

Page 78

Name _____ Date _____

Find the missing numbers. Solid lines mean multiply.
Dotted lines mean add.

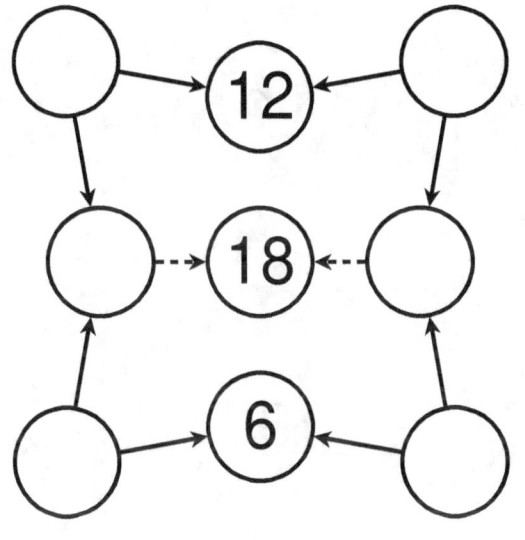

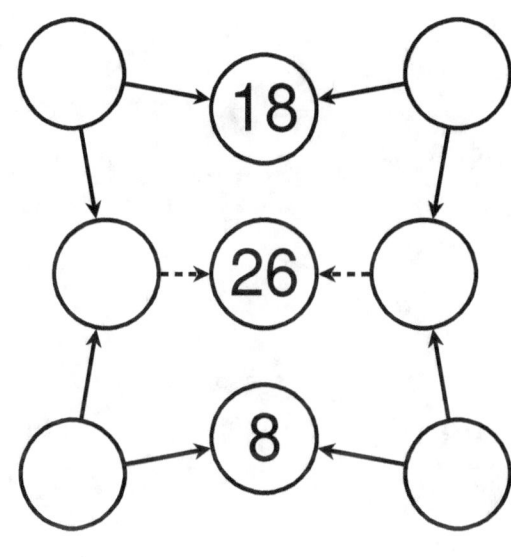

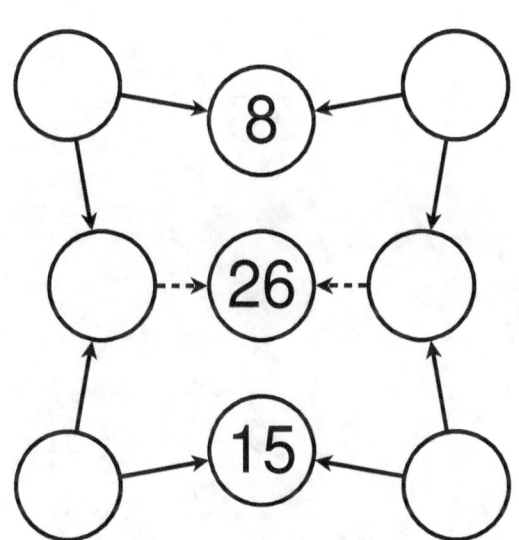

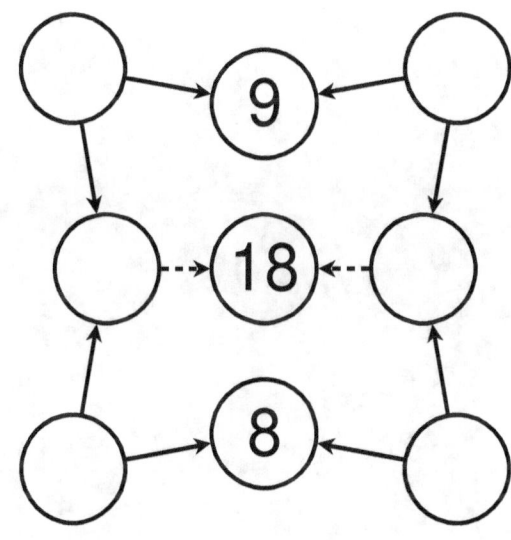

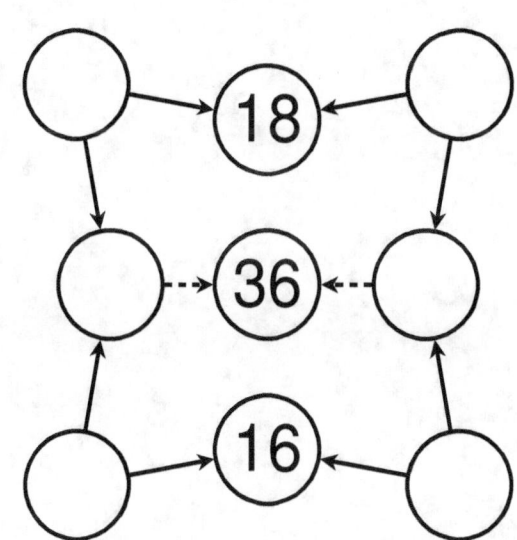

Name _____   Date _____

Find the missing numbers. Solid lines mean multiply.
Dotted lines mean add.

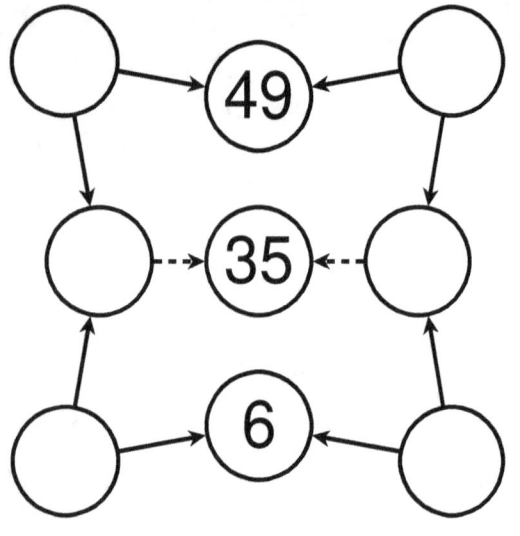

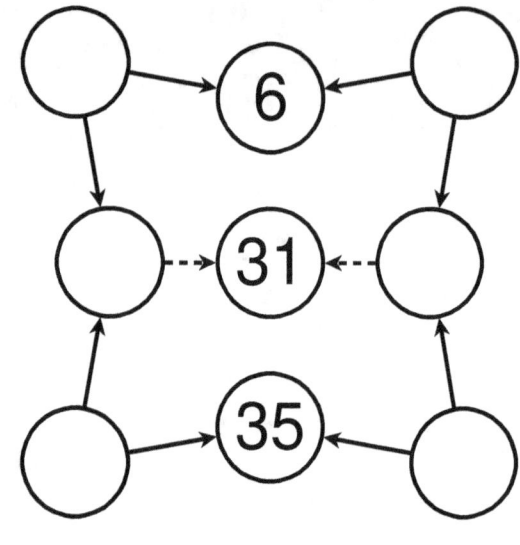

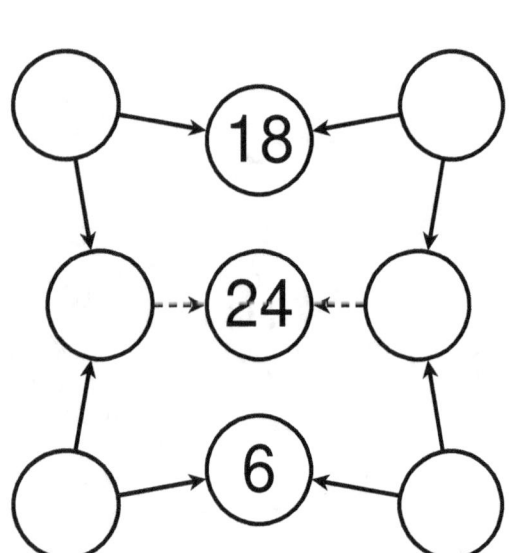

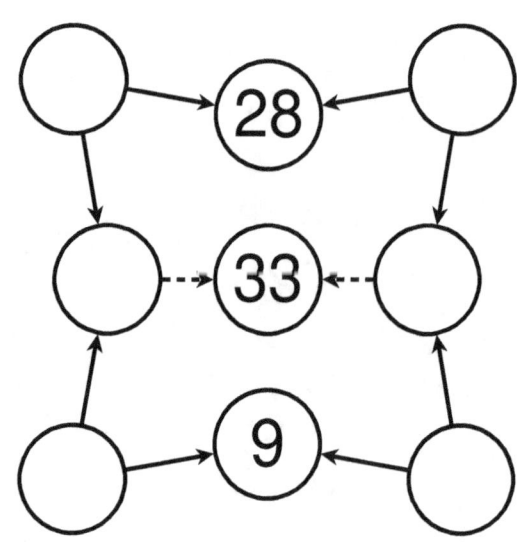

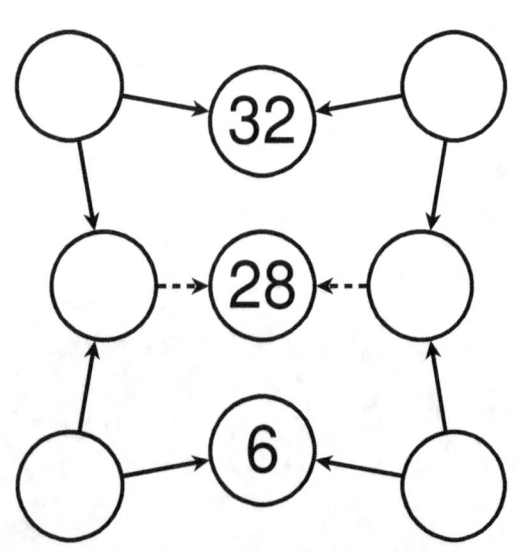

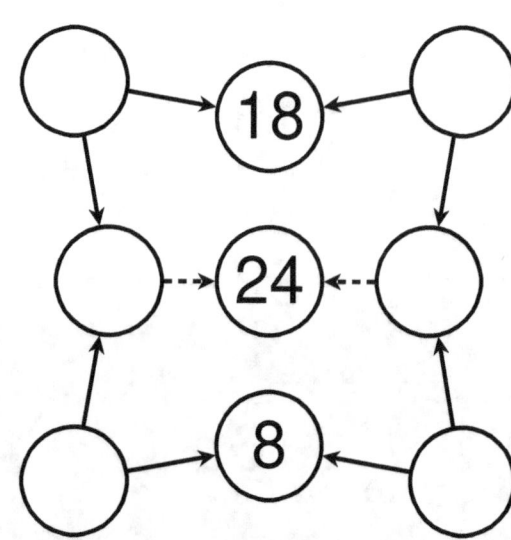

Page 80

Name _____   Date _____

Find the missing numbers. Solid lines mean multiply.
Dotted lines mean add.

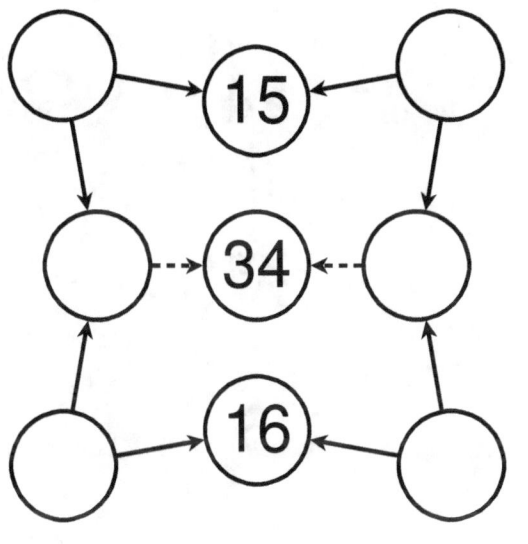

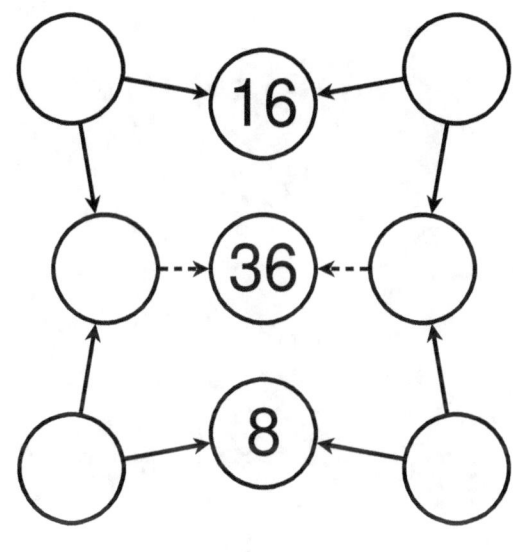

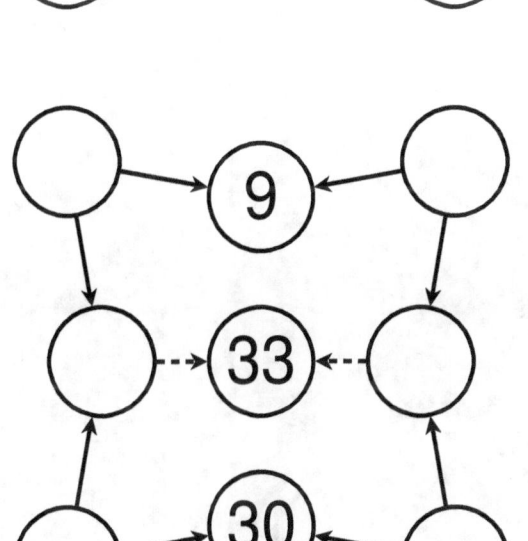

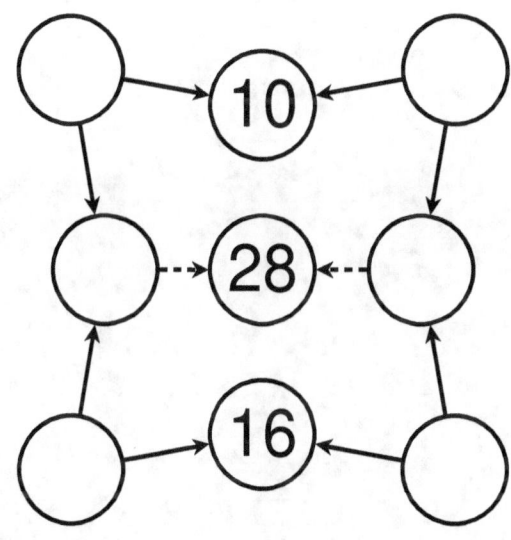

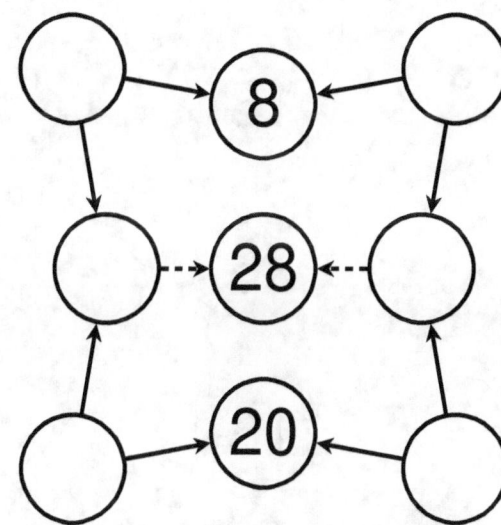

Name _____ Date _____

Find the missing numbers. Solid lines mean multiply.
Dotted lines mean add.

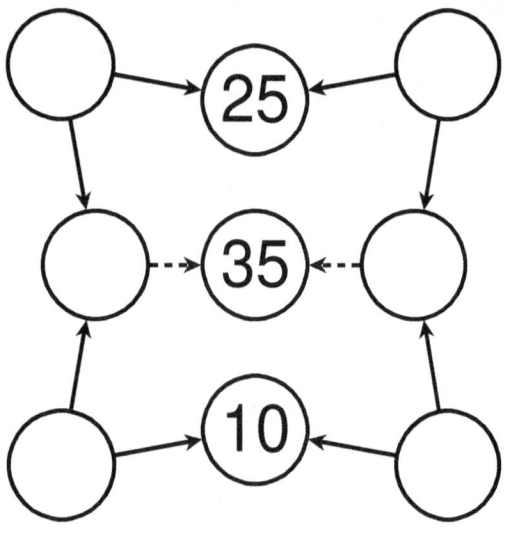

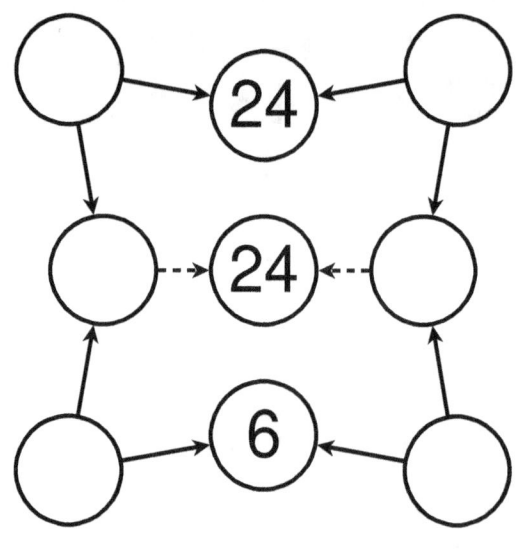

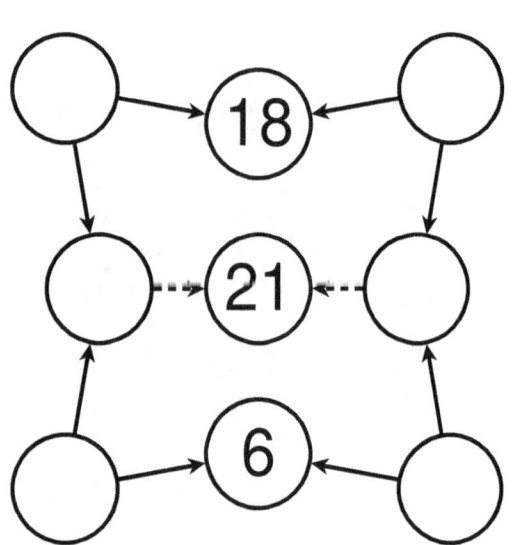

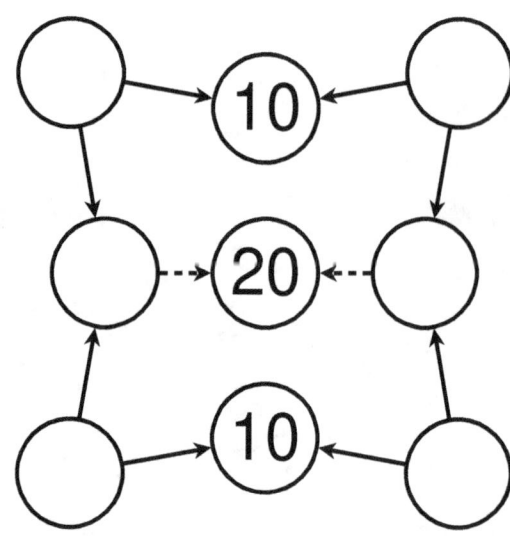

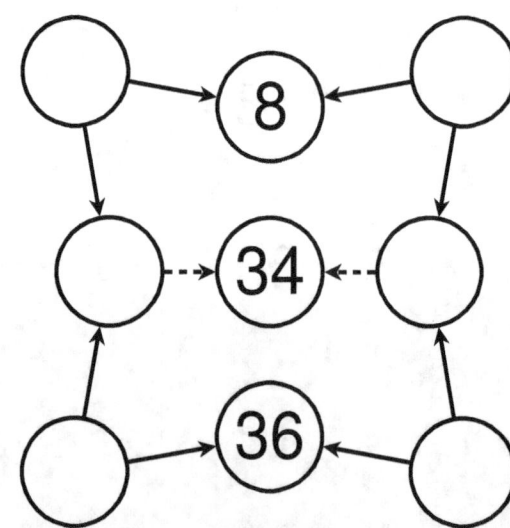

Name _____ Date _____

Find the missing numbers. Solid lines mean multiply.
Dotted lines mean add.

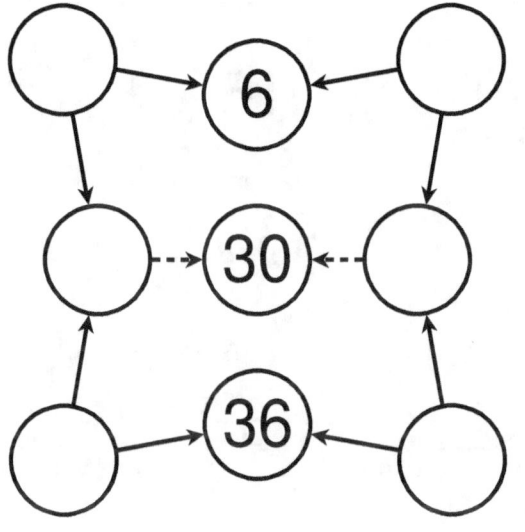

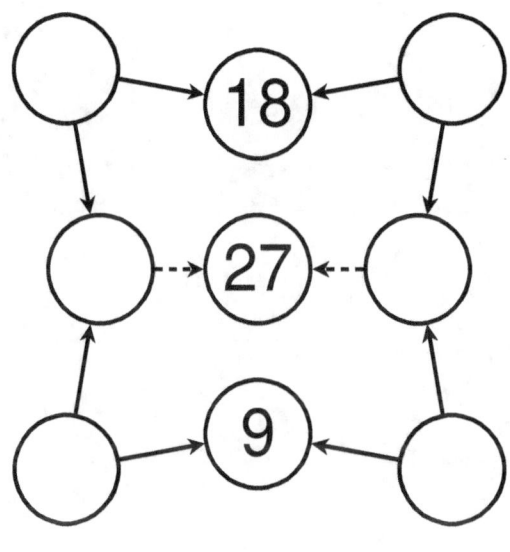

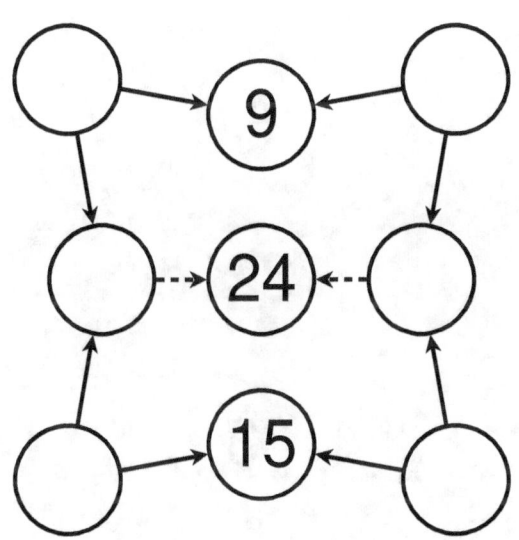

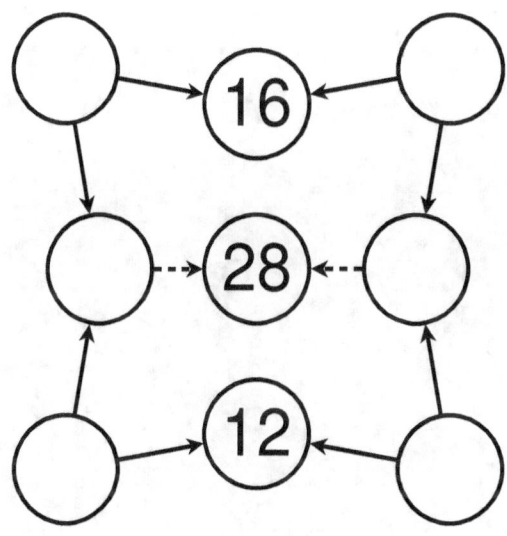

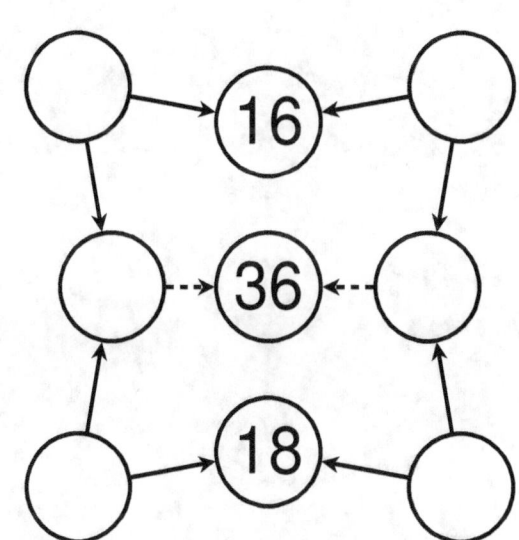

Name _____ Date _____

Find the missing numbers. Solid lines mean multiply.
Dotted lines mean add.

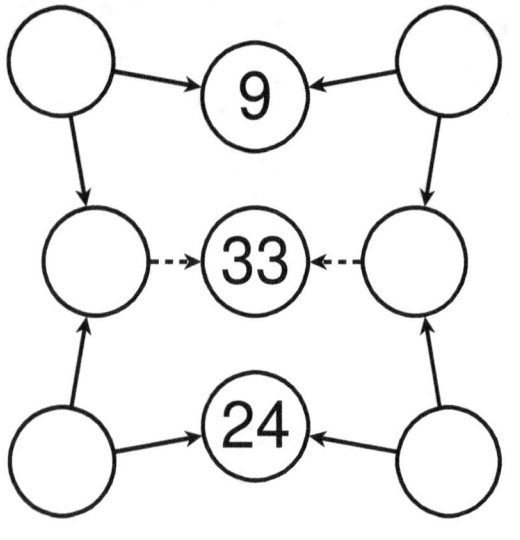

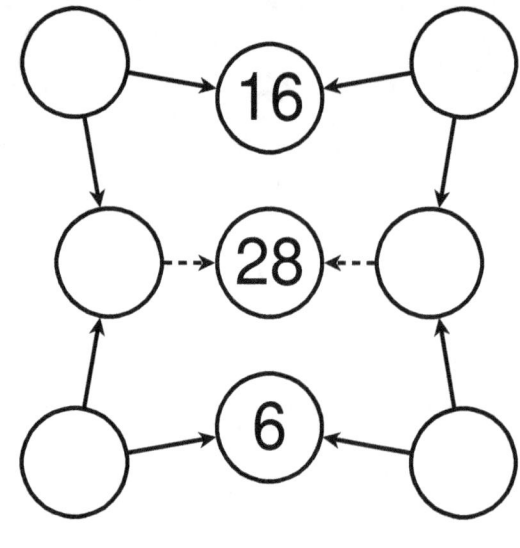

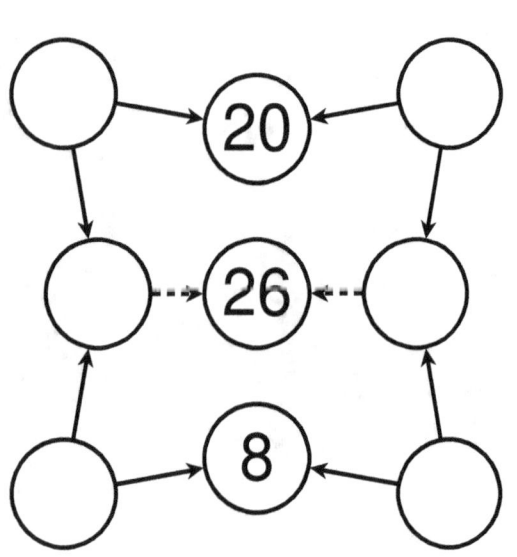

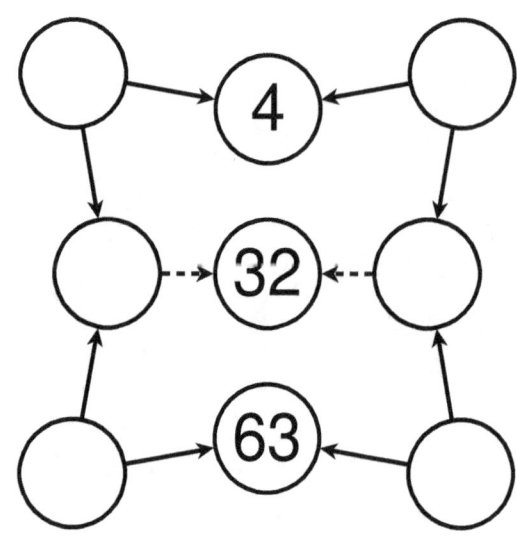

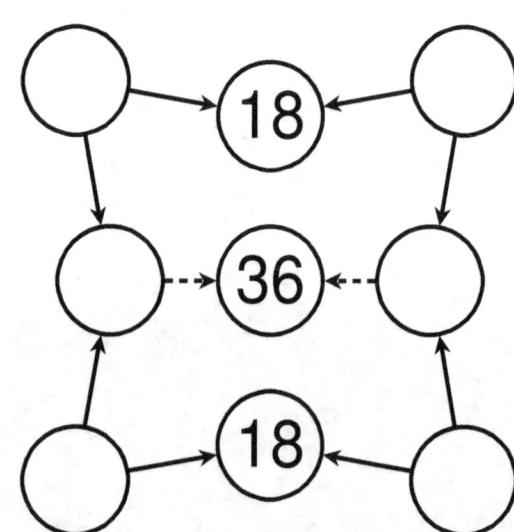

Name _____   Date _____

Find the missing numbers. Solid lines mean multiply.
Dotted lines mean add.

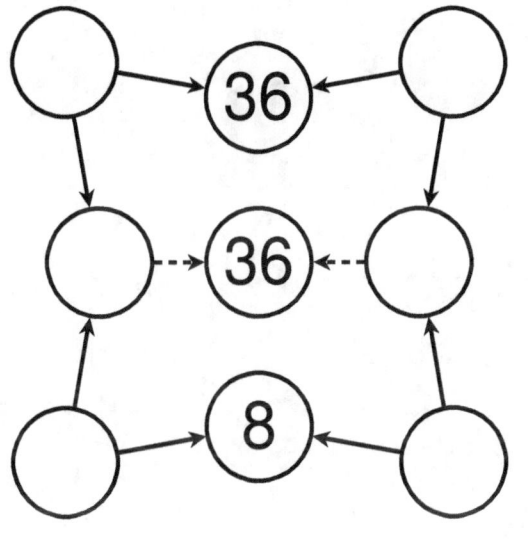

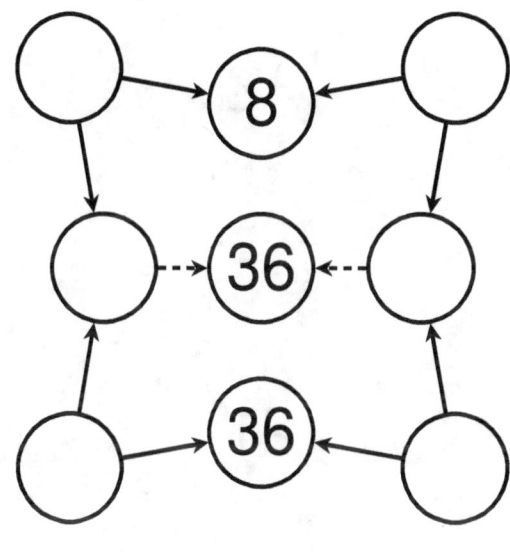

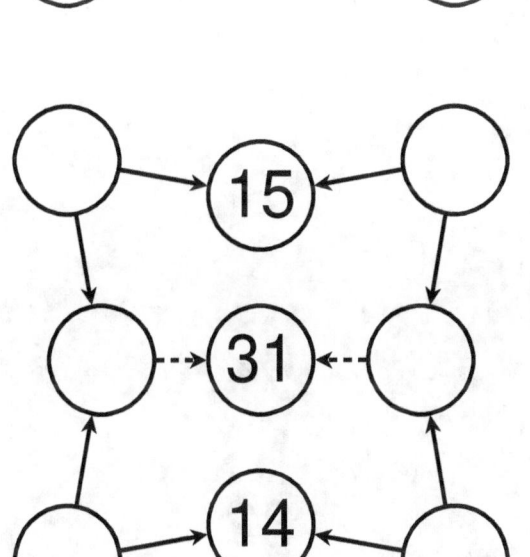

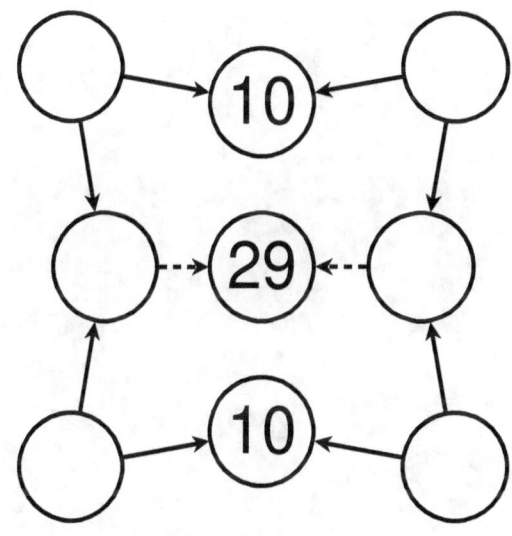

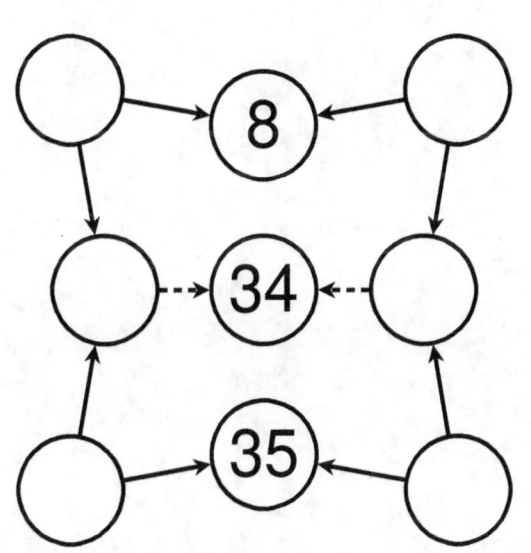

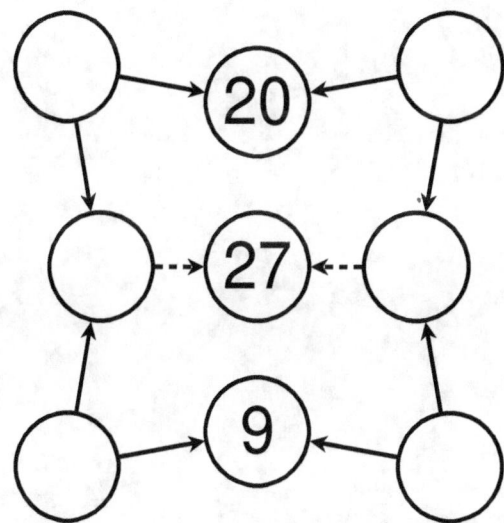

Name _____ Date _____

Find the missing numbers. Solid lines mean multiply.
Dotted lines mean add.

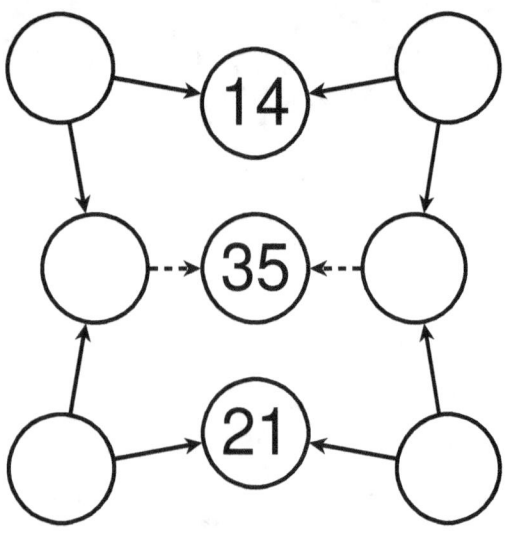

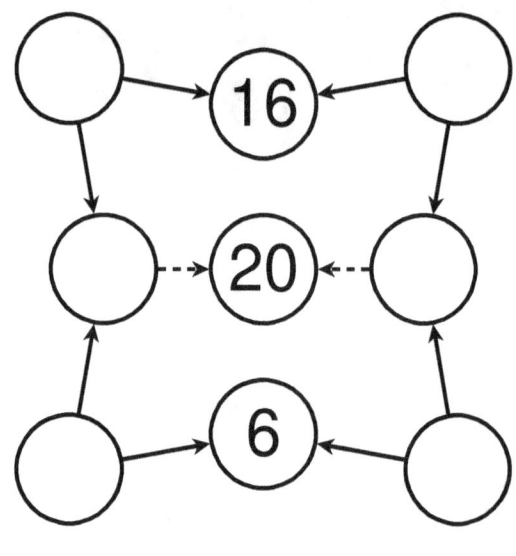

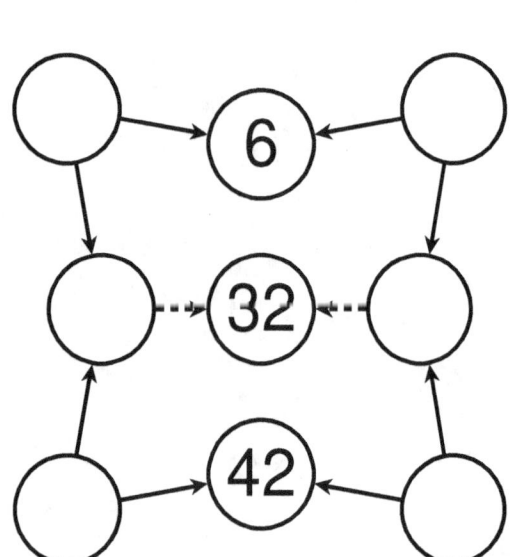

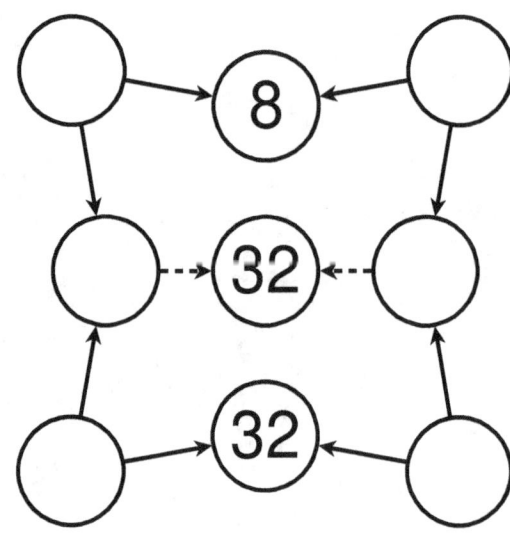

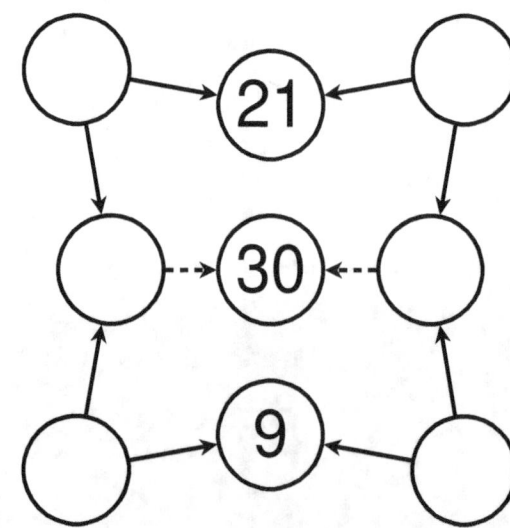

Name _____ Date _____

Find the missing numbers. Solid lines mean multiply.
Dotted lines mean add.

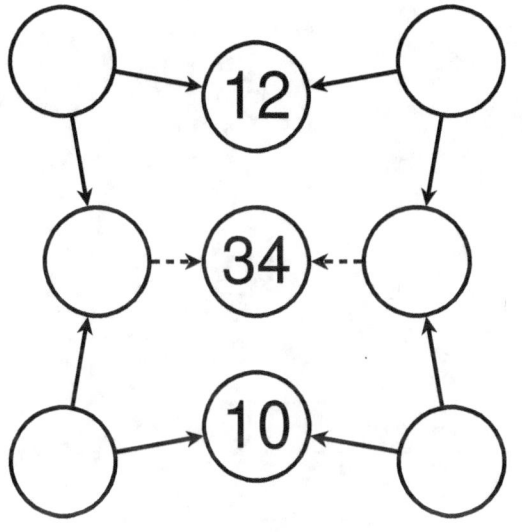

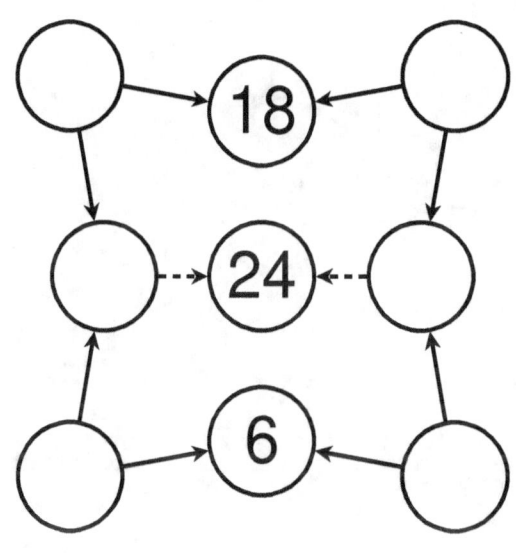

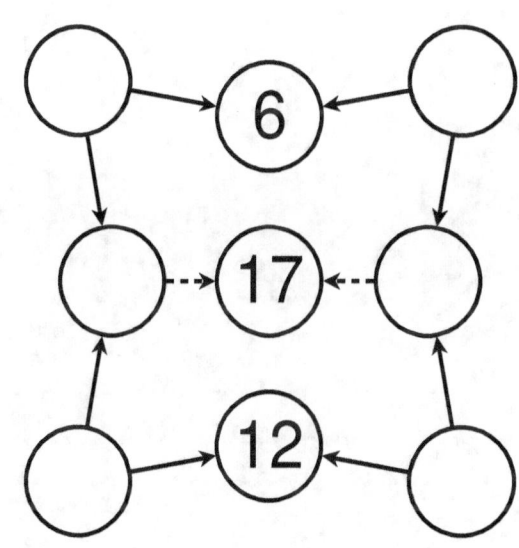

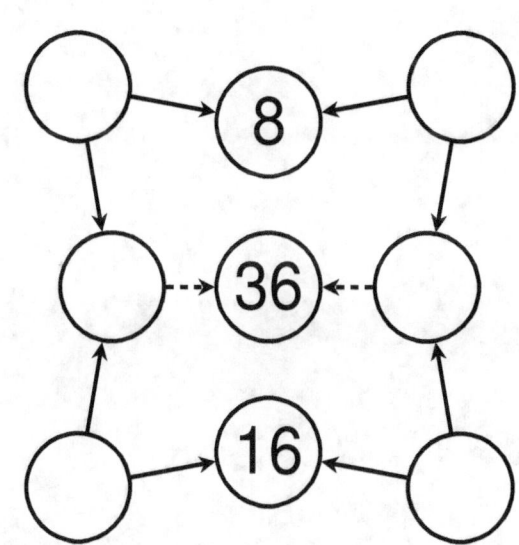

Name _____ Date _____

Find the missing numbers. Solid lines mean multiply.
Dotted lines mean add.

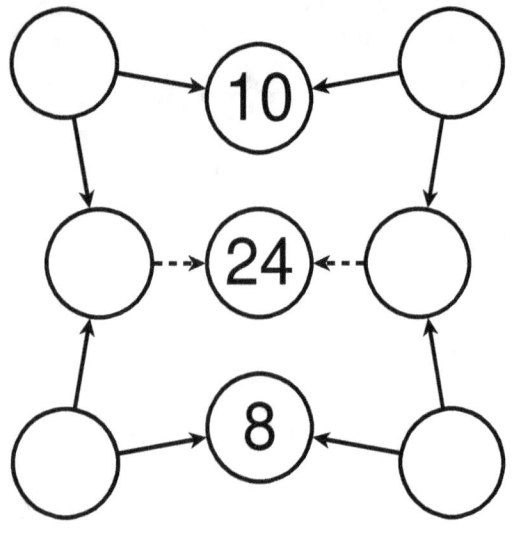

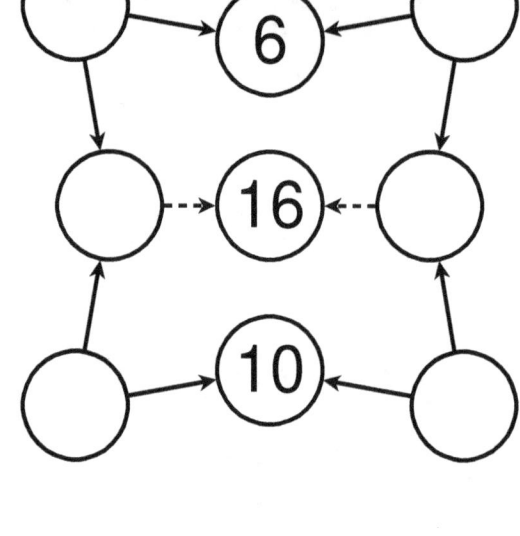

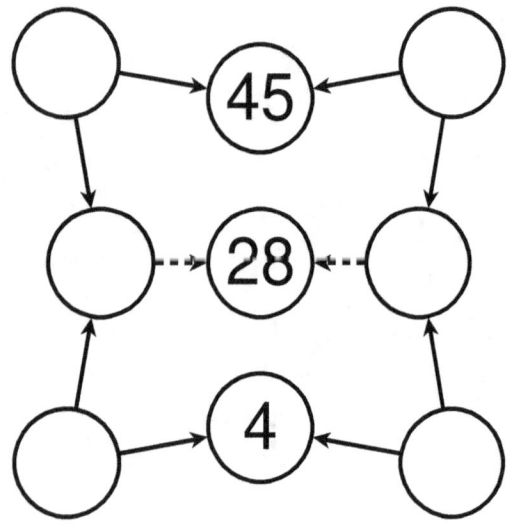

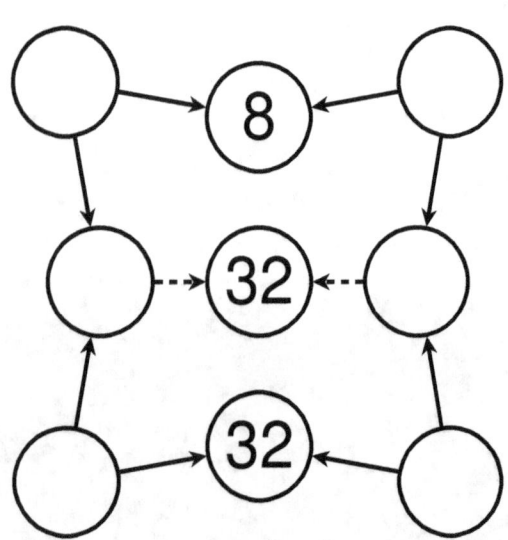

Name _____ Date _____

Find the missing numbers. Solid lines mean multiply.
Dotted lines mean add.

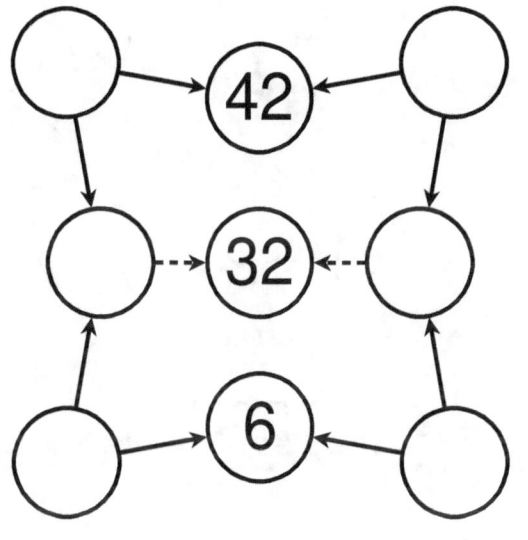

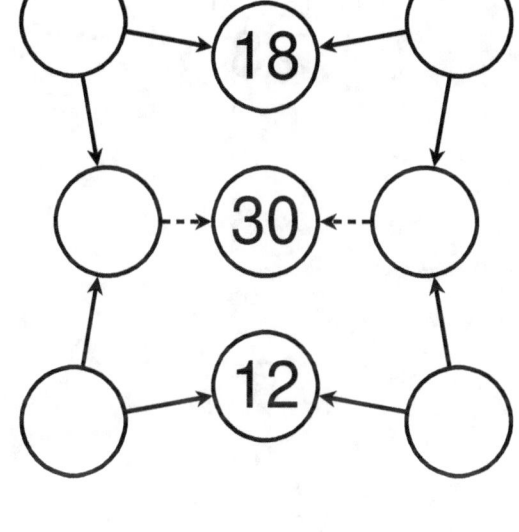

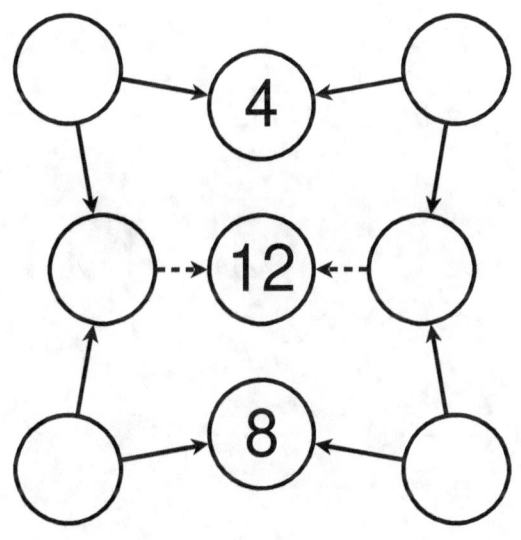

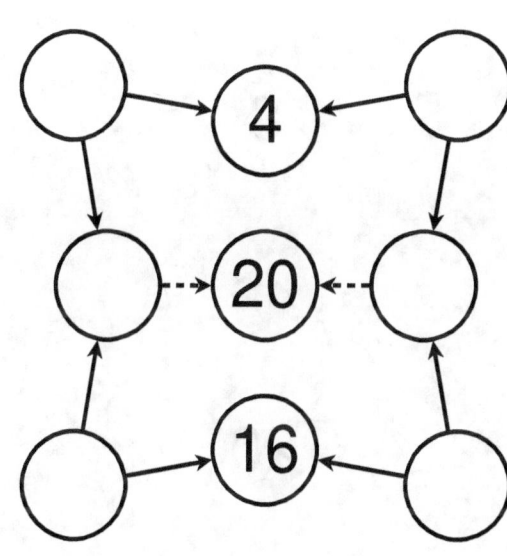

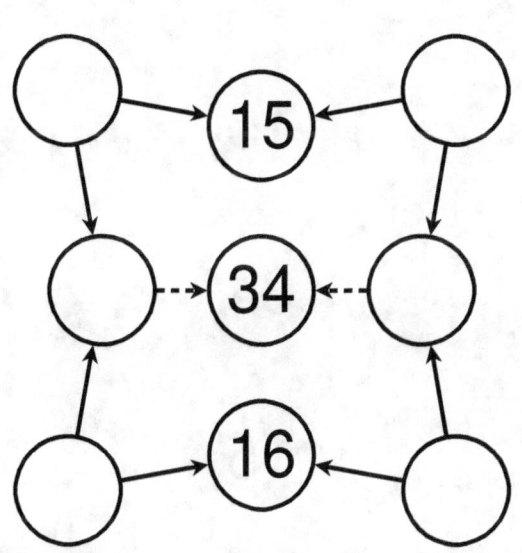

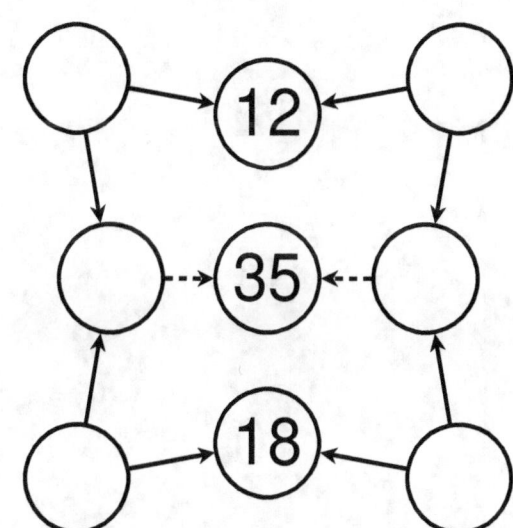

Name _____   Date _____

Find the missing numbers. Solid lines mean multiply.
Dotted lines mean add.

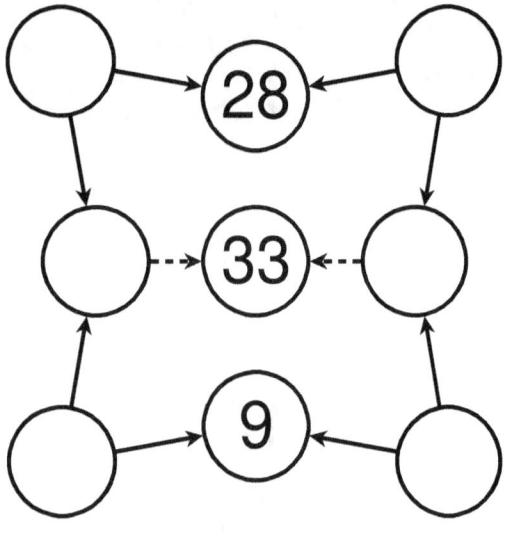

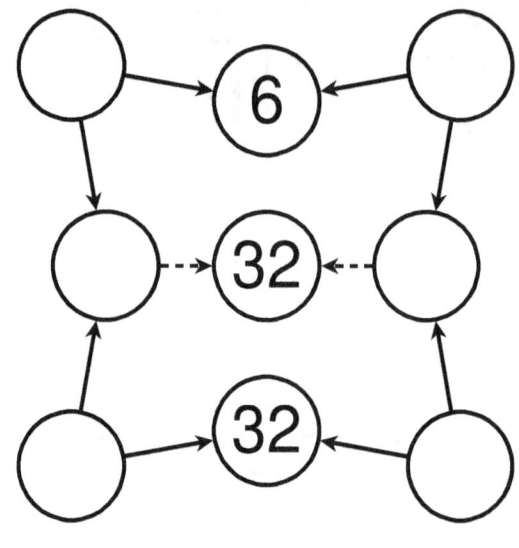

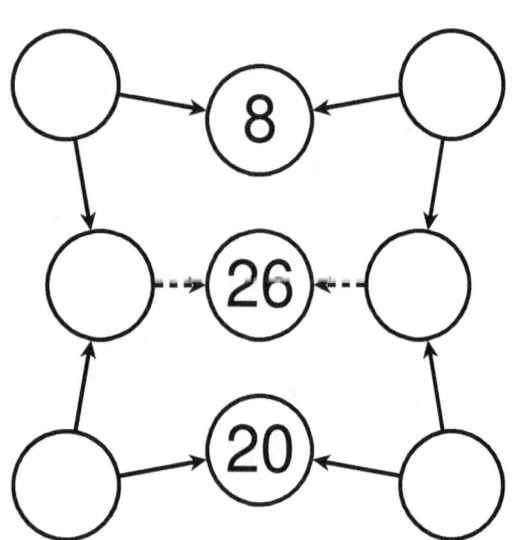

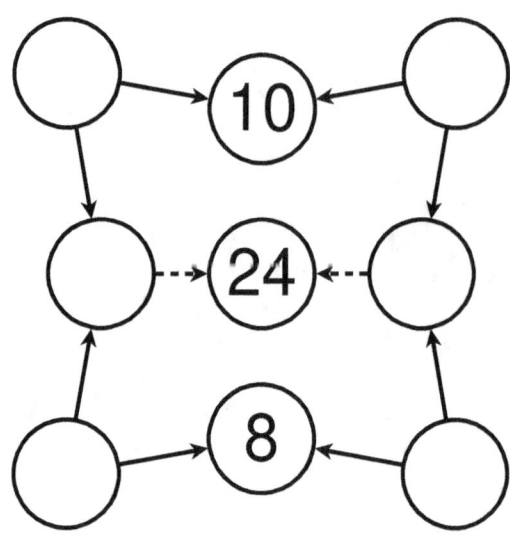

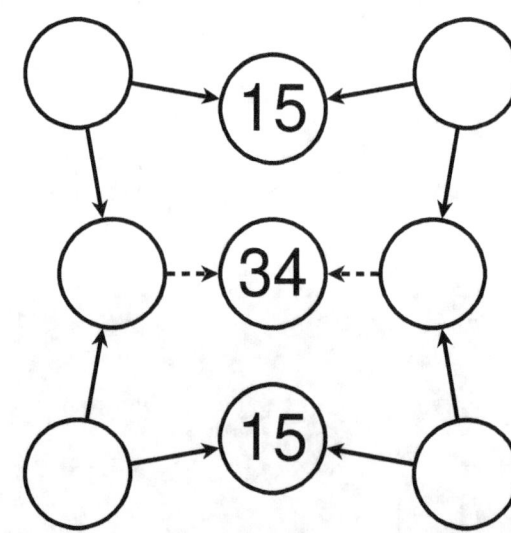

Page 90

Name _____ Date _____

Find the missing numbers. Solid lines mean multiply.
Dotted lines mean add.

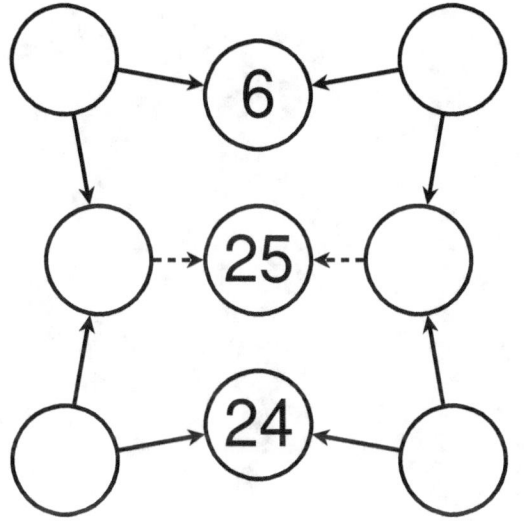

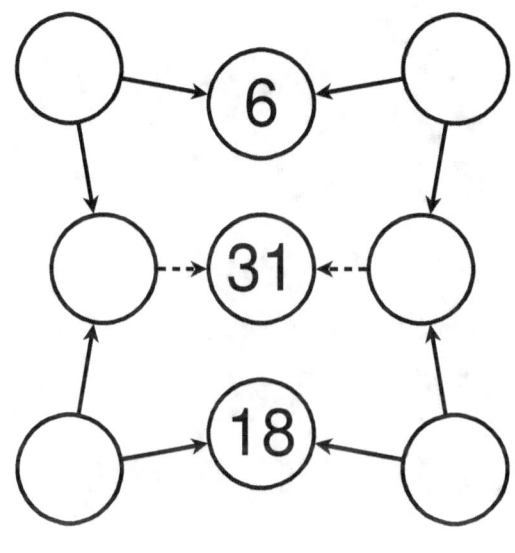

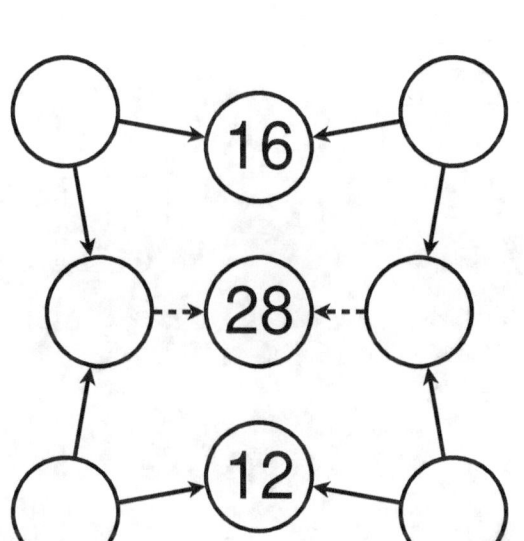

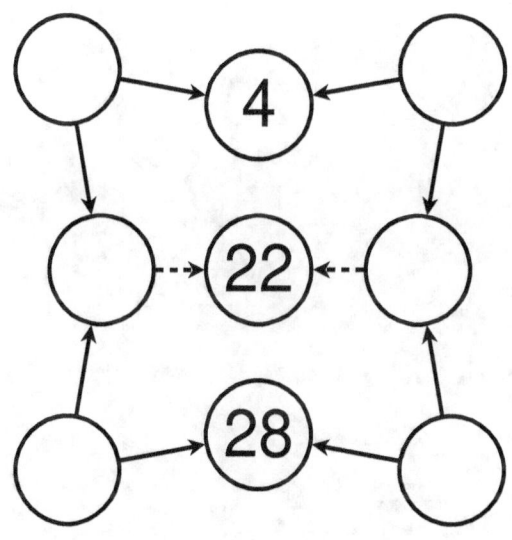

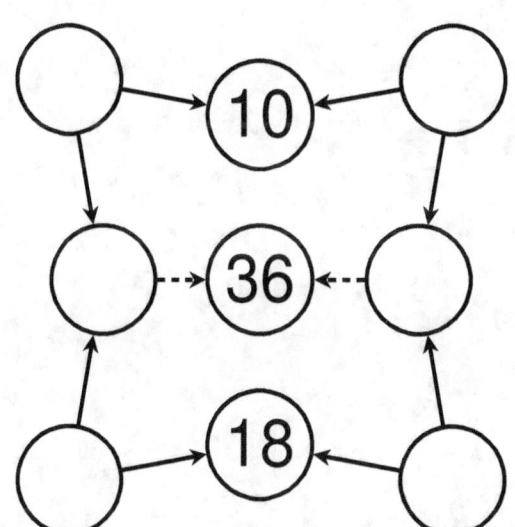

Name _____ Date _____

Find the missing numbers. Solid lines mean multiply.
Dotted lines mean add.

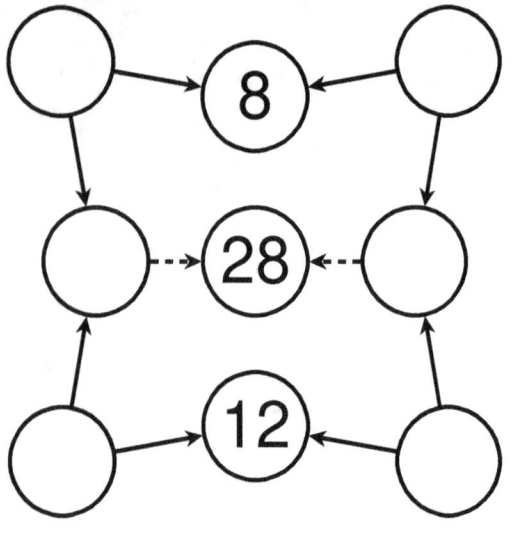

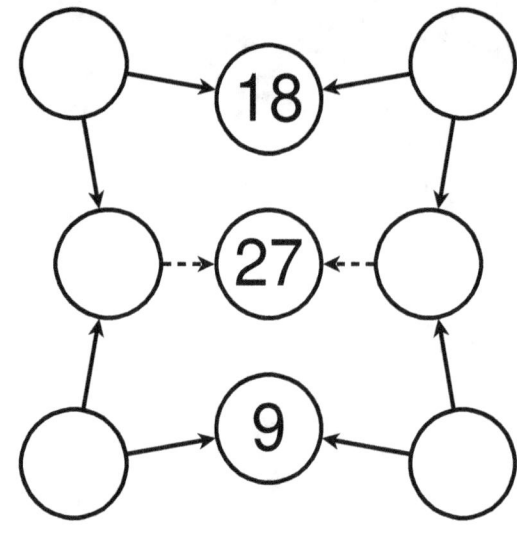

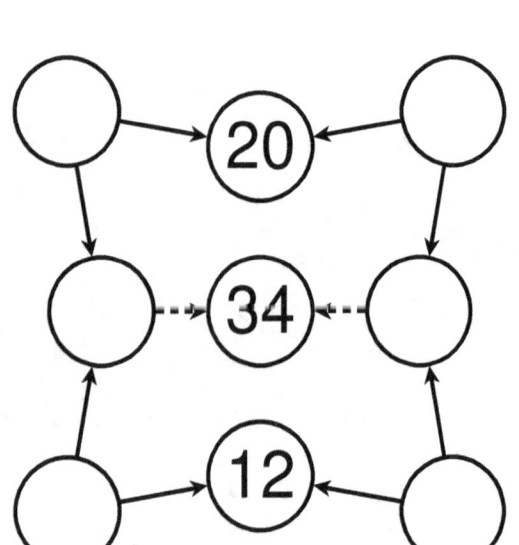

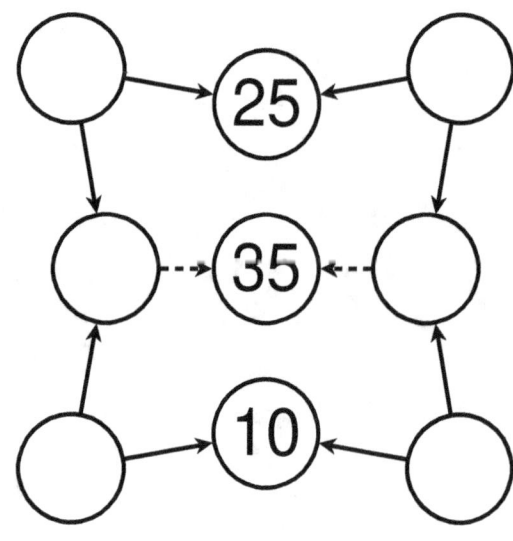

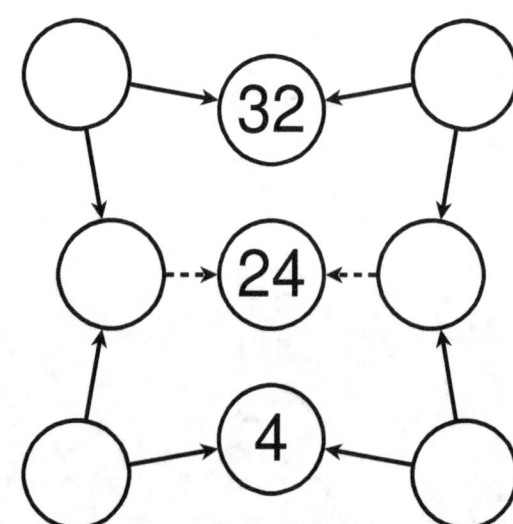

Name _____ Date _____

Find the missing numbers. Solid lines mean multiply.
Dotted lines mean add.

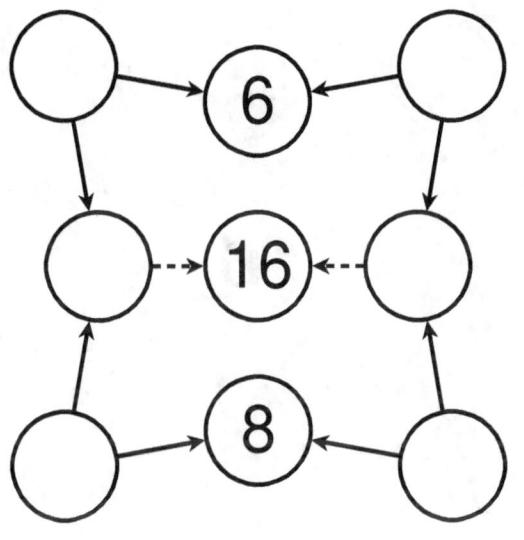

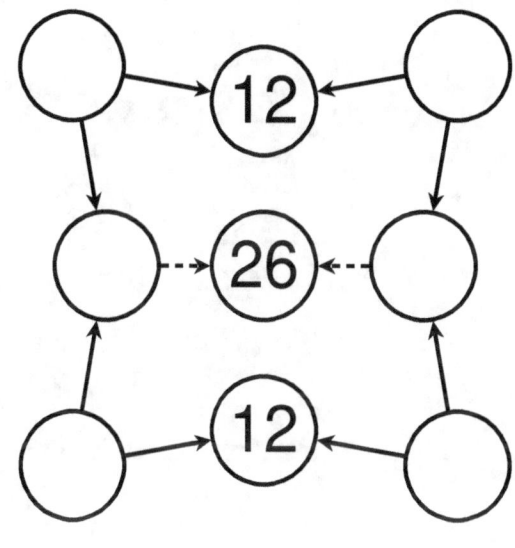

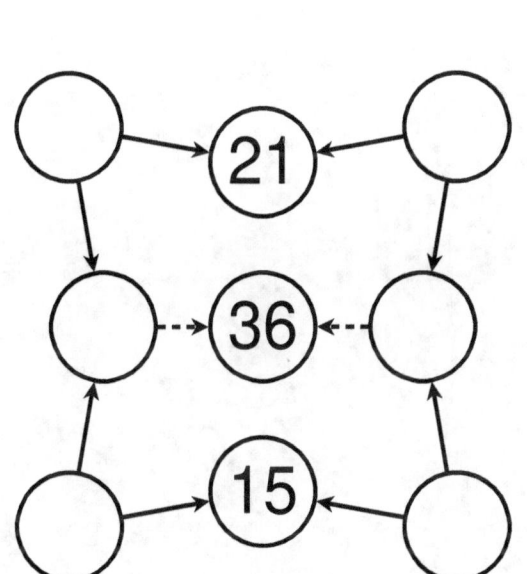

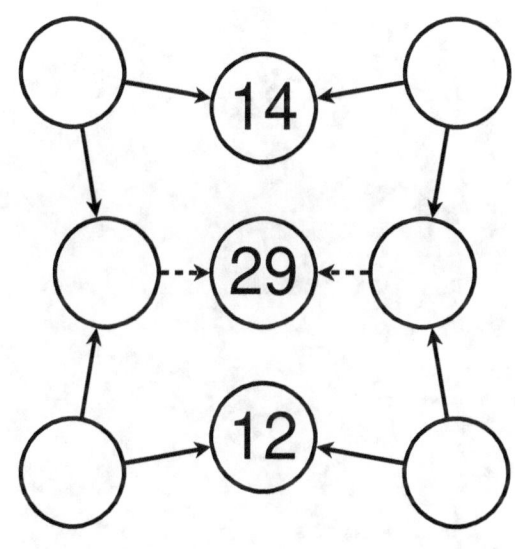

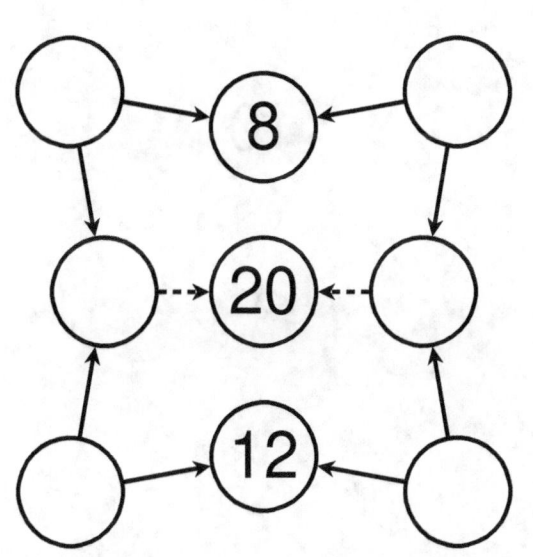

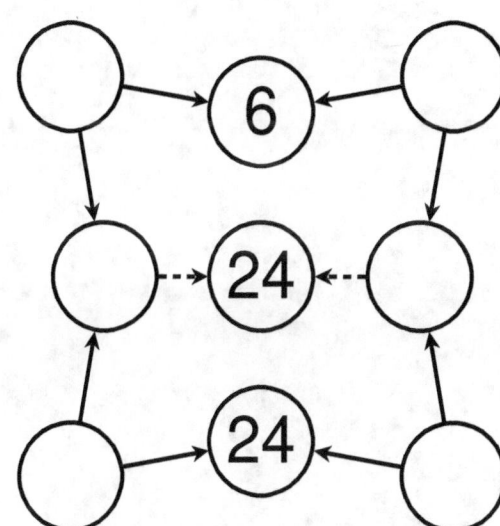

Name _____ Date _____

Find the missing numbers. Solid lines mean multiply.
Dotted lines mean add.

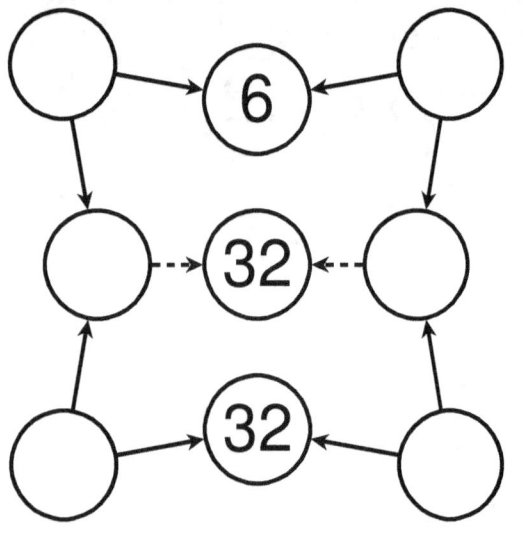

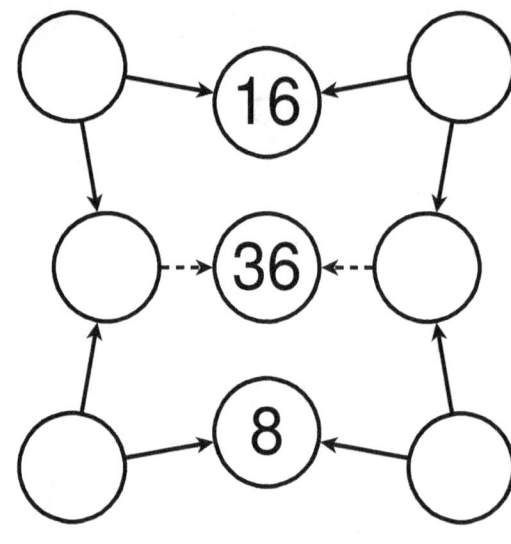

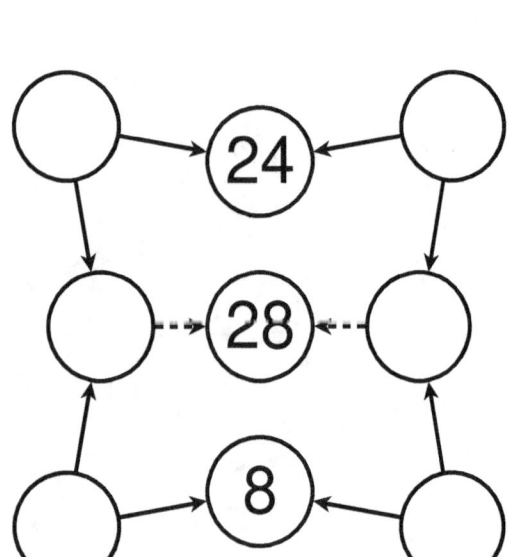

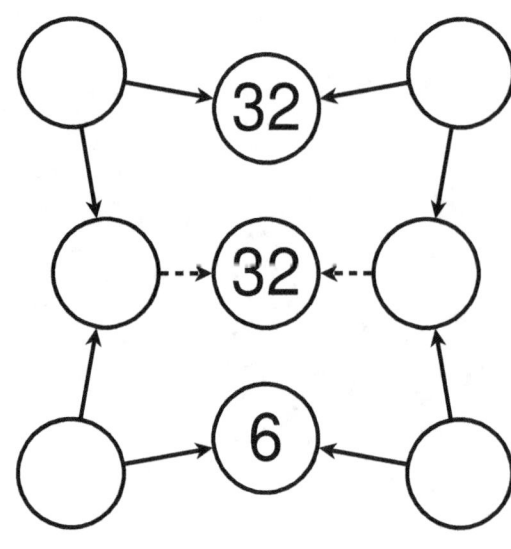

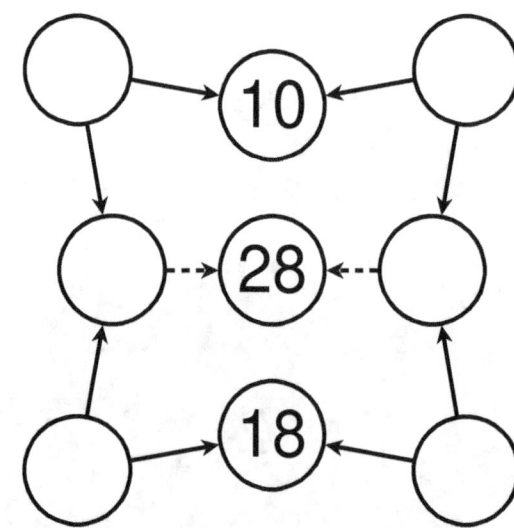

Page 94

Find the missing numbers. Solid lines mean multiply.
Dotted lines mean add.

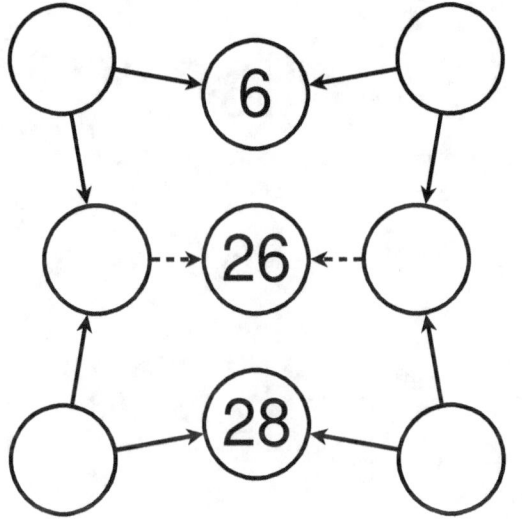

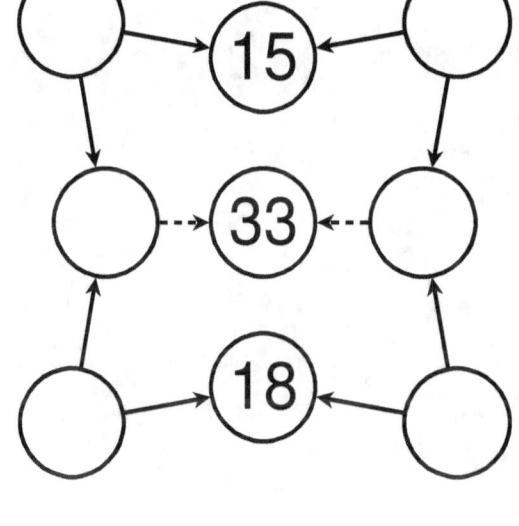

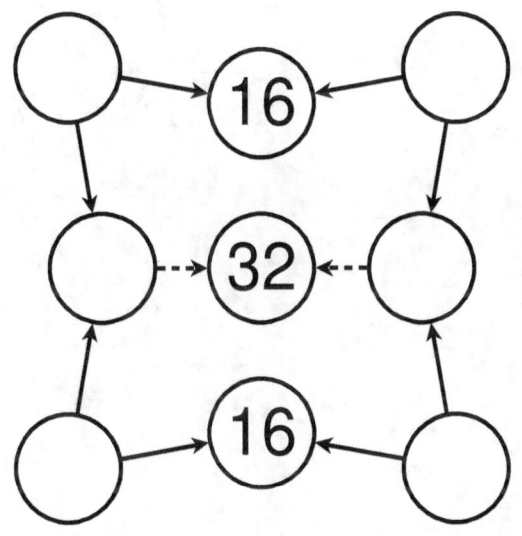

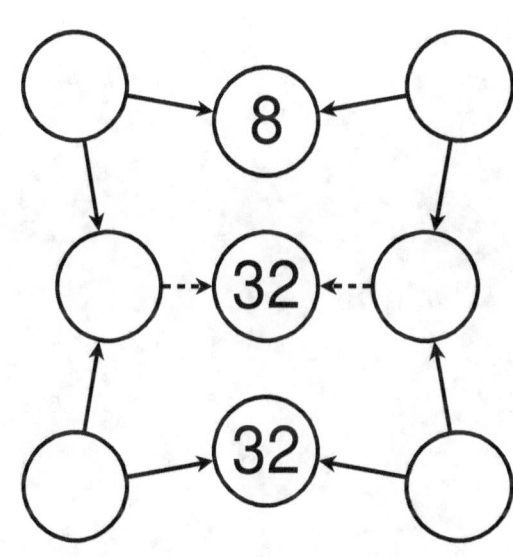

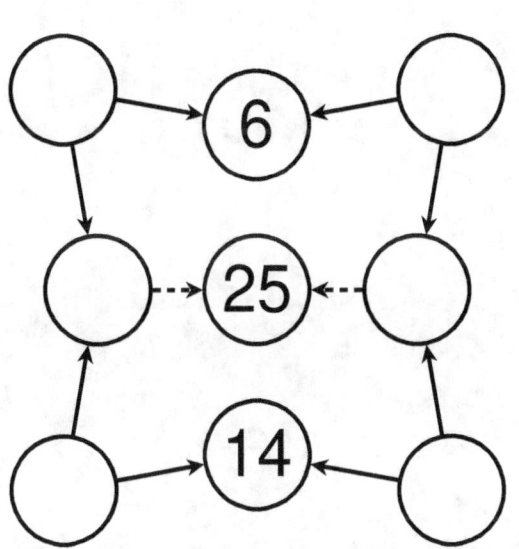

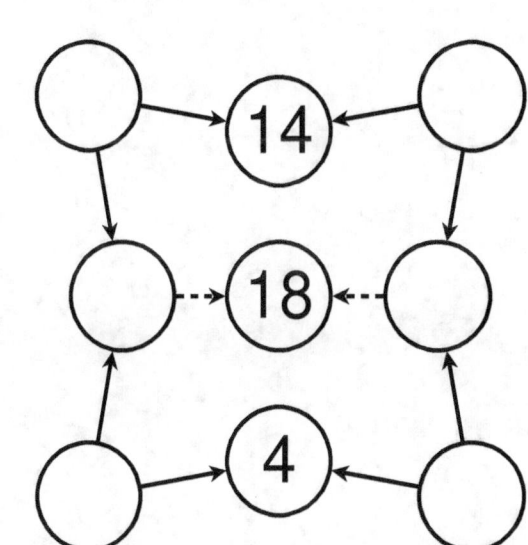

Name _____ Date _____

Find the missing numbers. Solid lines mean multiply.
Dotted lines mean add.

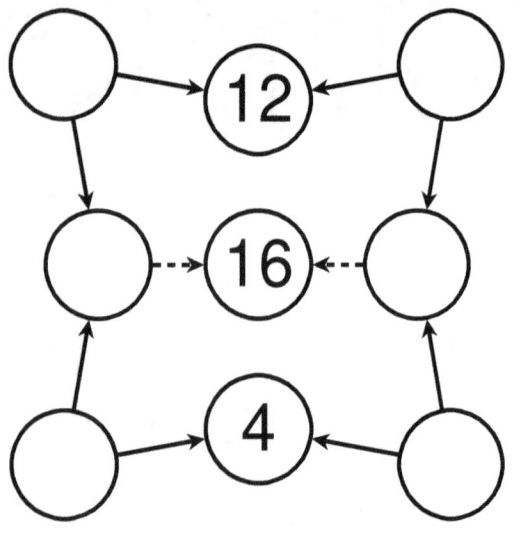

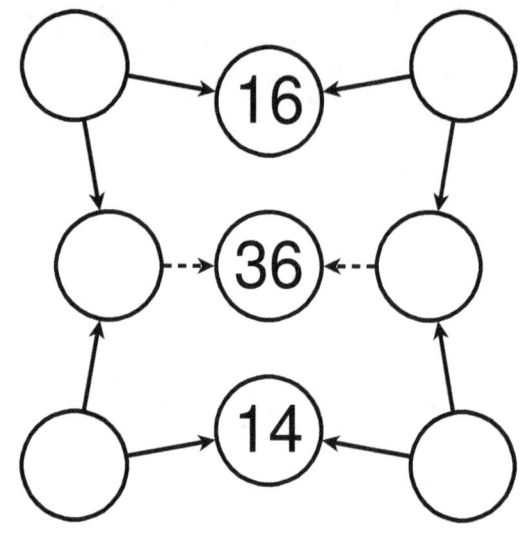

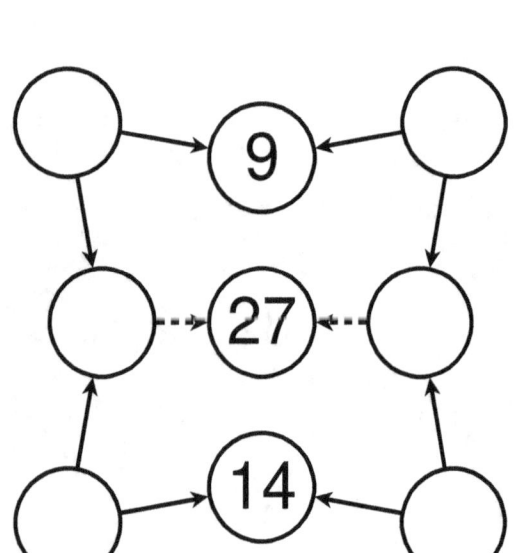

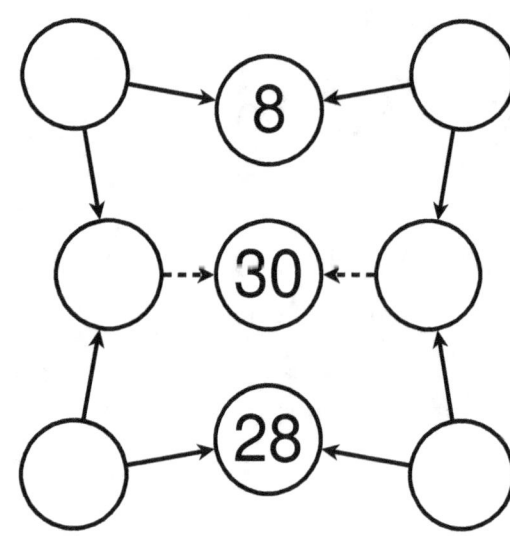

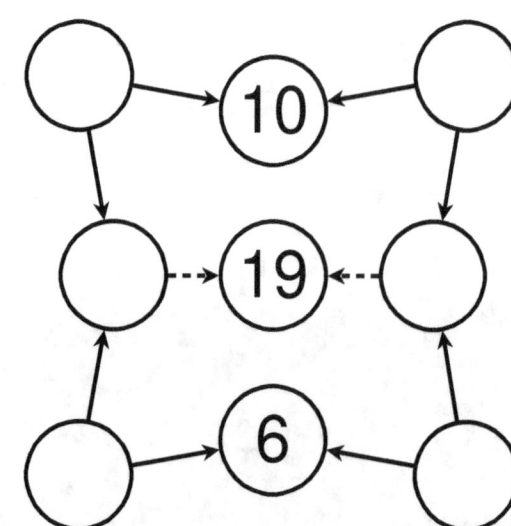

Name _____ Date _____

Find the missing numbers. Solid lines mean multiply.
Dotted lines mean add.

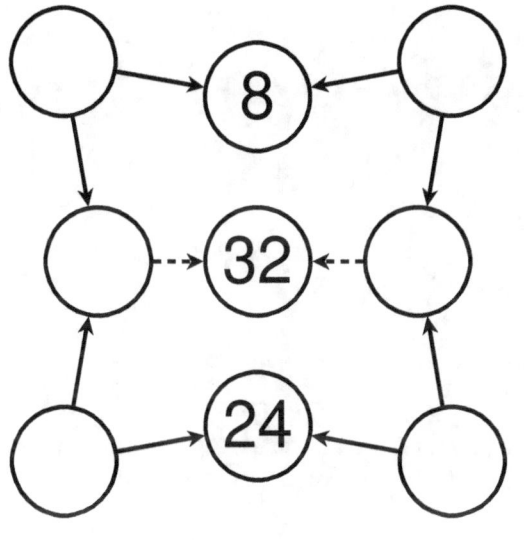

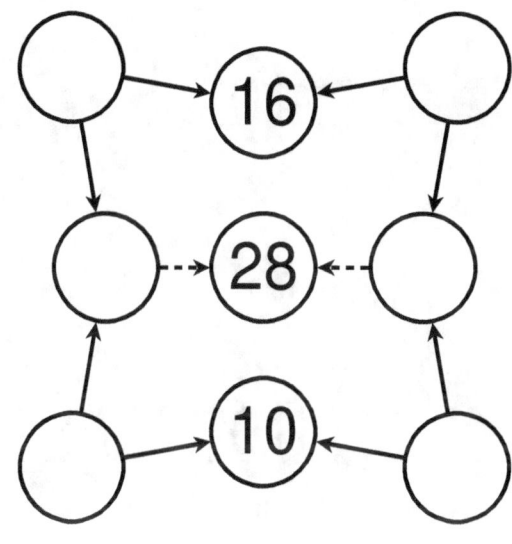

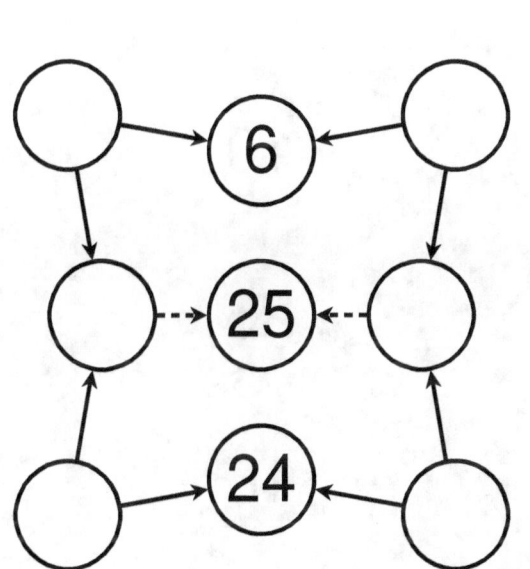

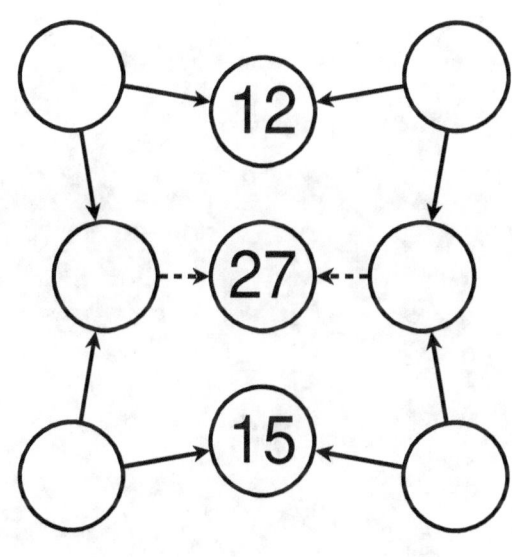

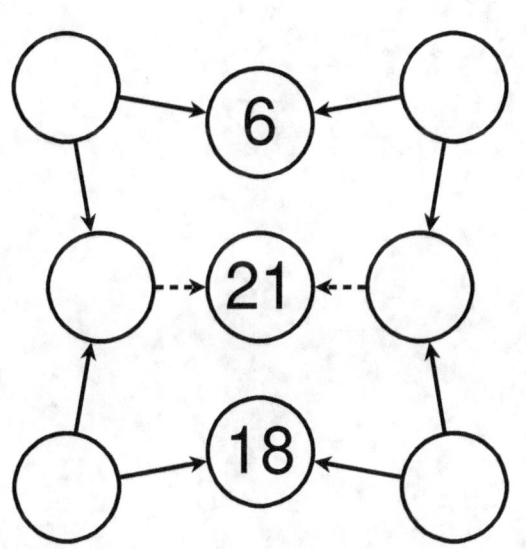

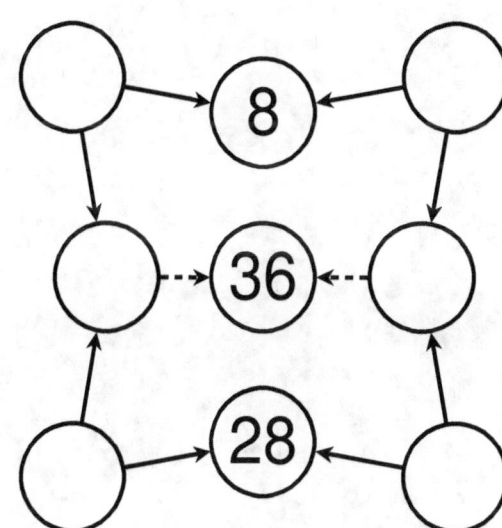

Name _____ Date _____

Find the missing numbers. Solid lines mean multiply.
Dotted lines mean add.

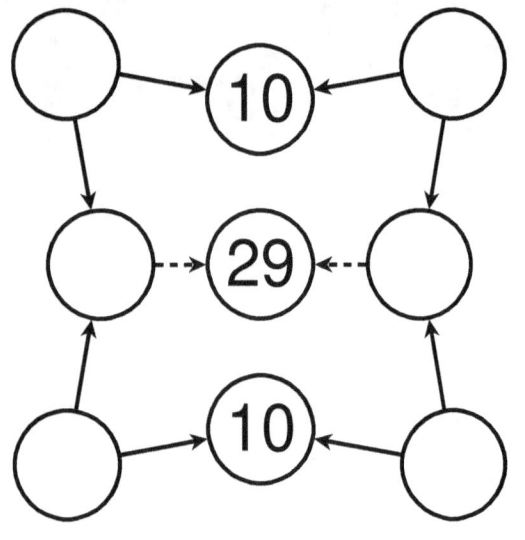

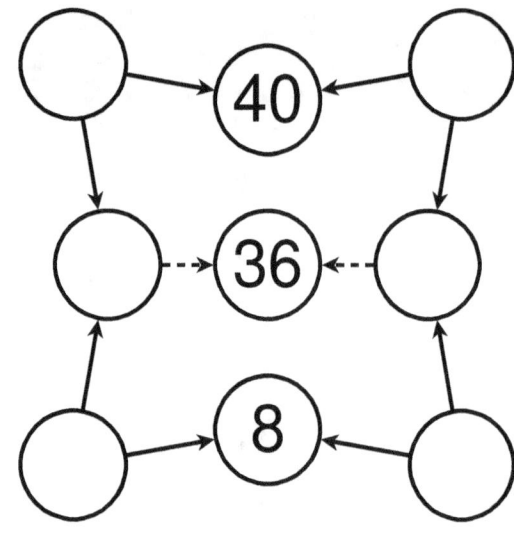

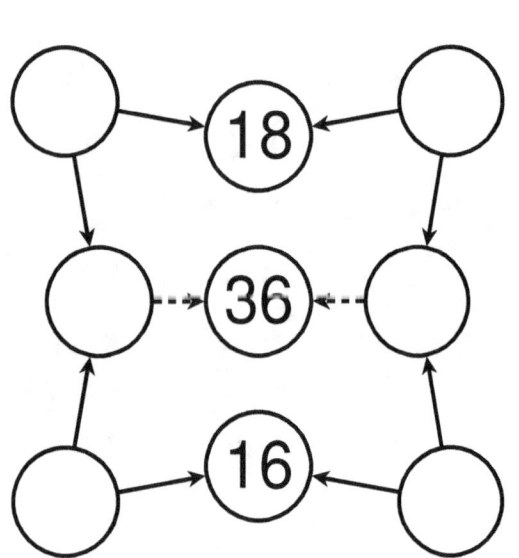

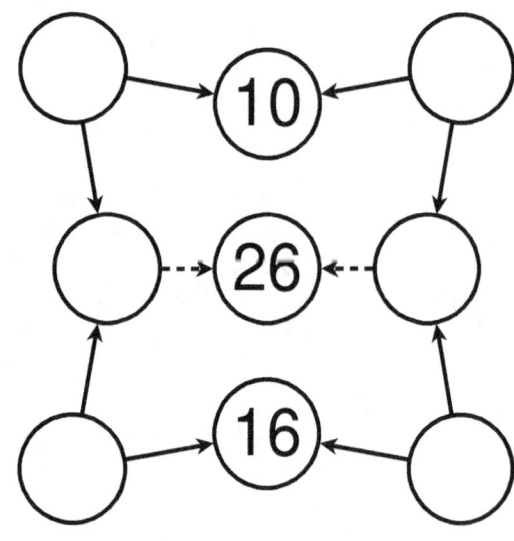

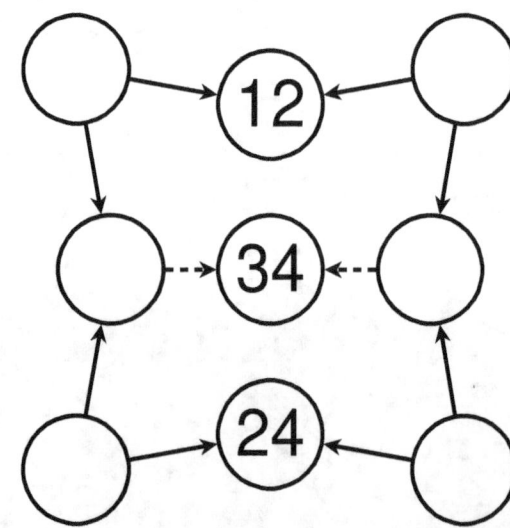

Page 98

Name _____ Date _____

Find the missing numbers. Solid lines mean multiply.
Dotted lines mean add.

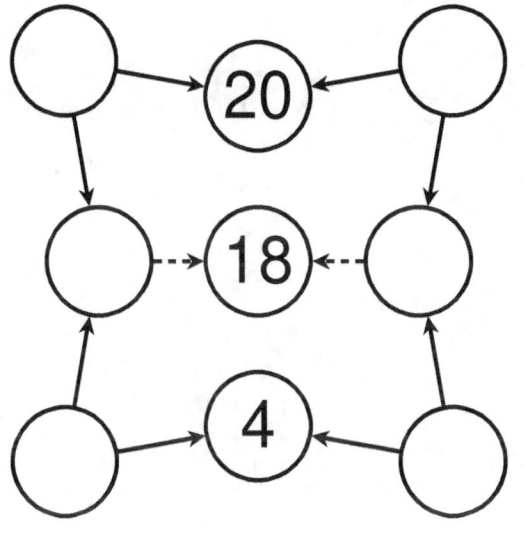

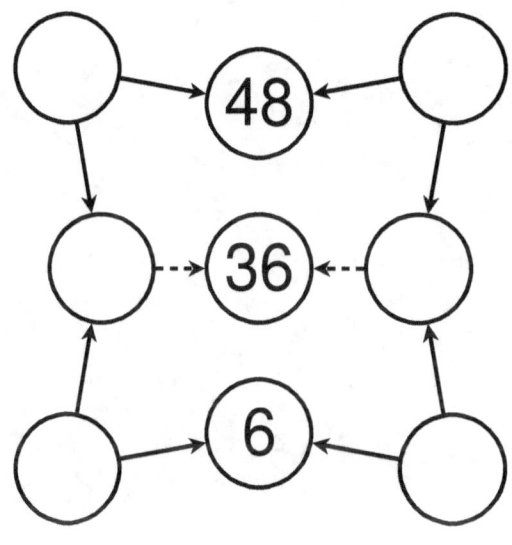

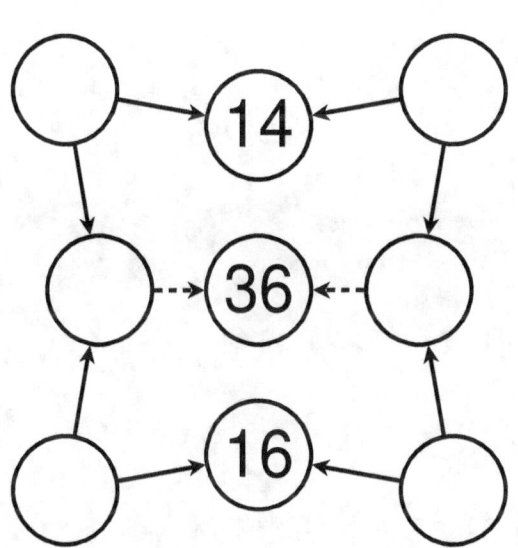

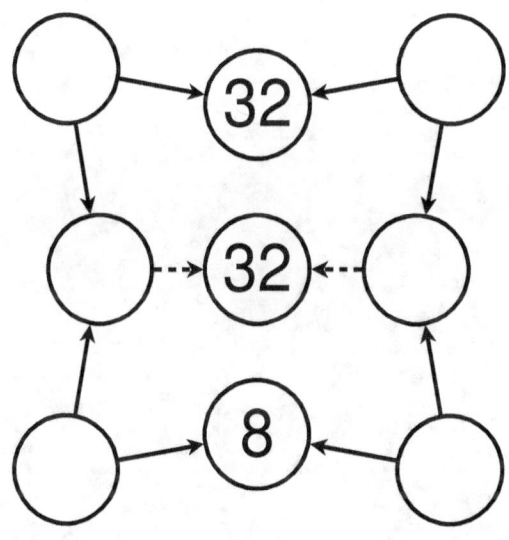

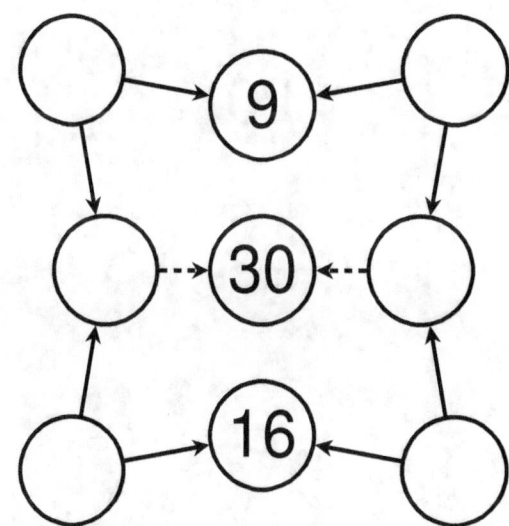

Name _____ Date _____

Find the missing numbers. Solid lines mean multiply.
Dotted lines mean add.

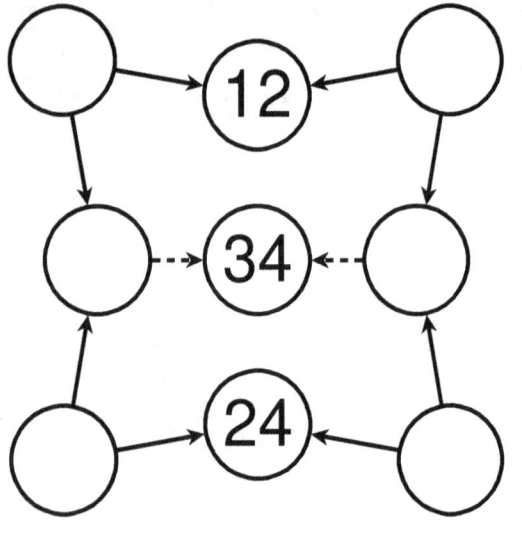

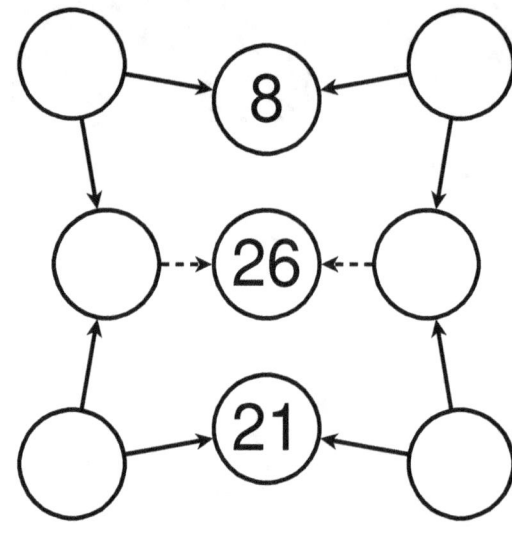

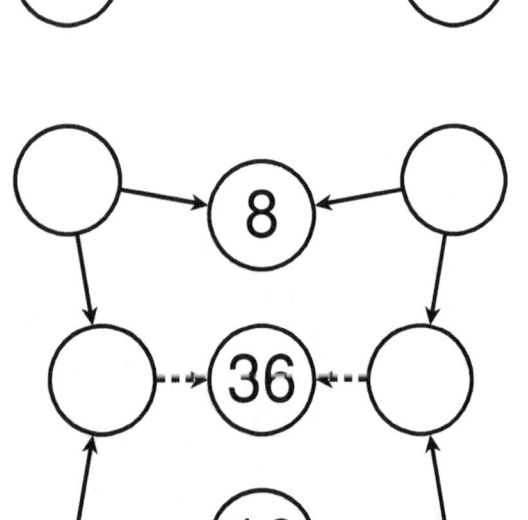

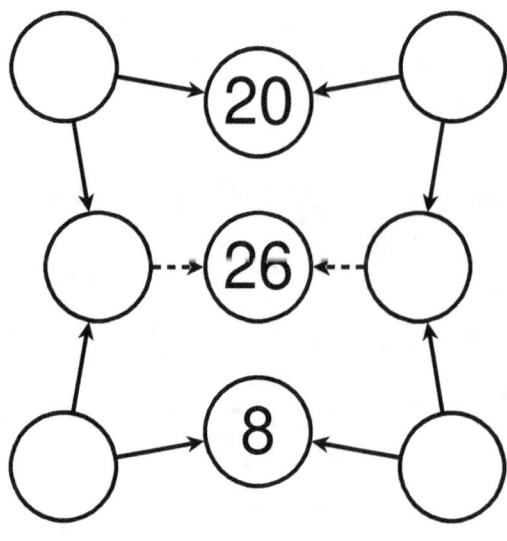

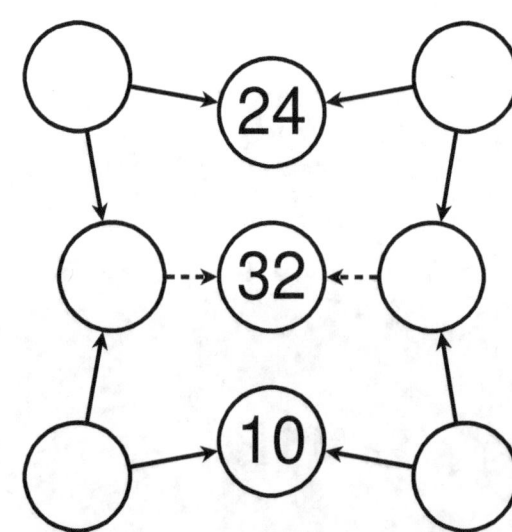

Solutions for Page 1

Find the missing numbers. Solid lines mean multiply.
Dotted lines mean add.

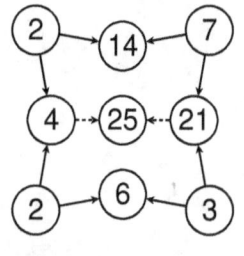
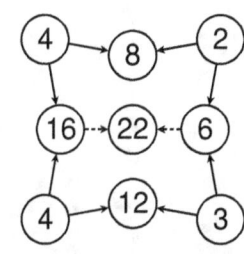

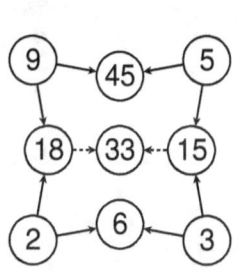
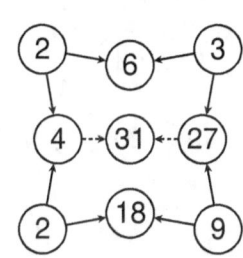

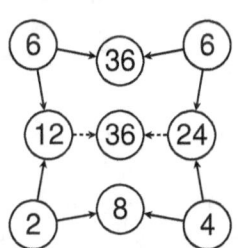
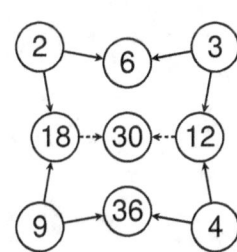

Solutions for Page 2

Find the missing numbers. Solid lines mean multiply.
Dotted lines mean add.

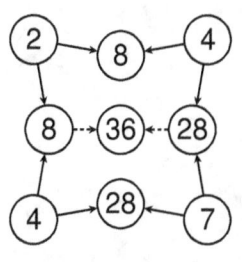
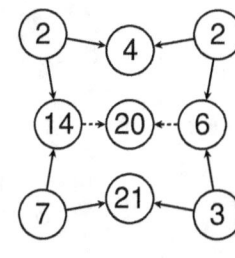

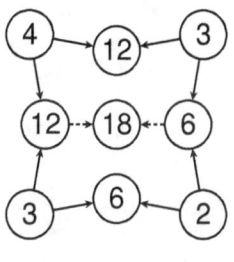
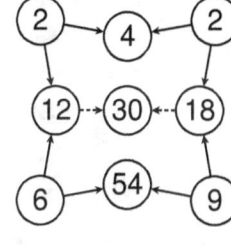

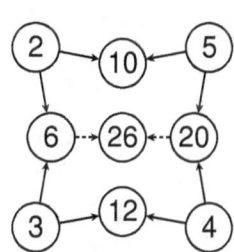
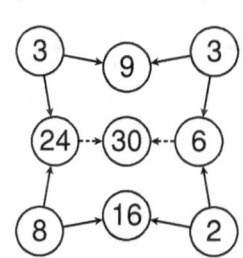

Solutions for Page 3

Find the missing numbers. Solid lines mean multiply.
Dotted lines mean add.

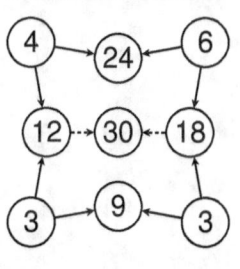
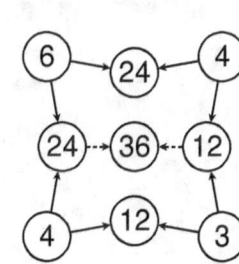

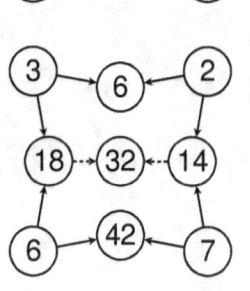
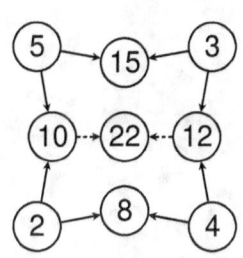

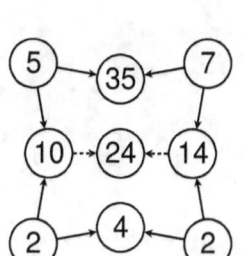
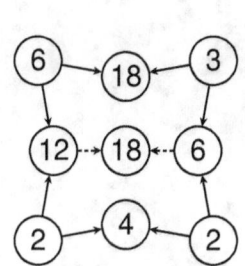

Solutions for Page 4

Find the missing numbers. Solid lines mean multiply.
Dotted lines mean add.

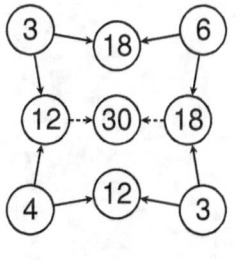
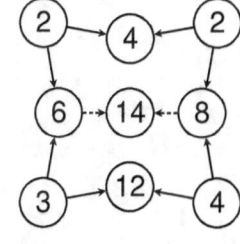

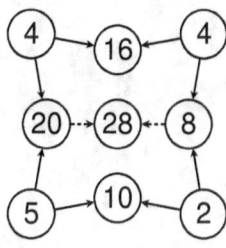
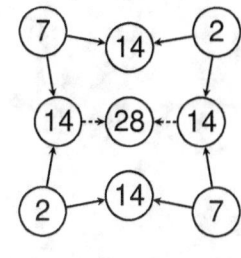

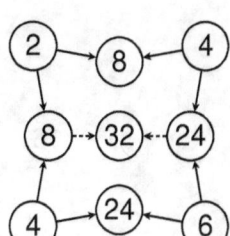
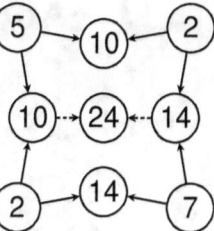

Solutions for Page 5

Find the missing numbers. Solid lines mean multiply. Dotted lines mean add.

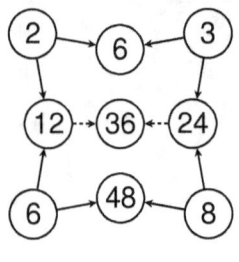 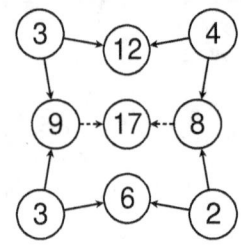

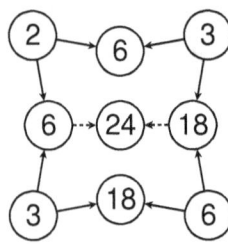 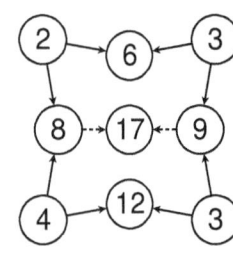

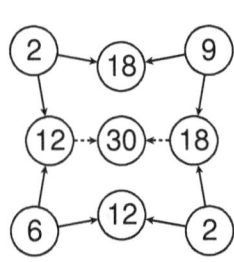 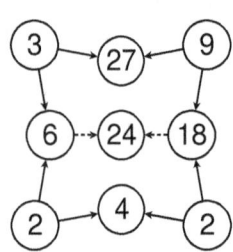

Solutions for Page 6

Find the missing numbers. Solid lines mean multiply. Dotted lines mean add.

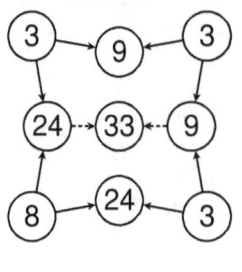 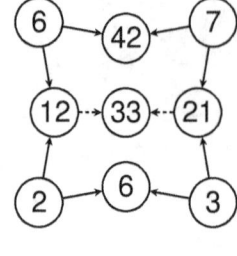

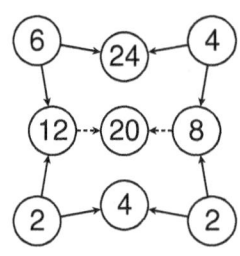 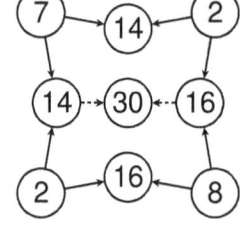

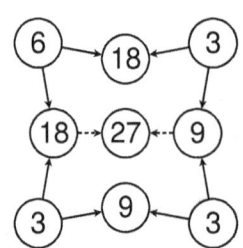 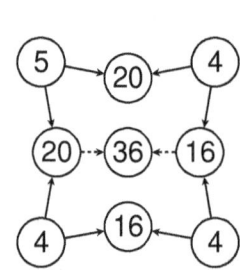

Solutions for Page 7

Find the missing numbers. Solid lines mean multiply. Dotted lines mean add.

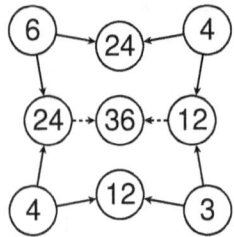 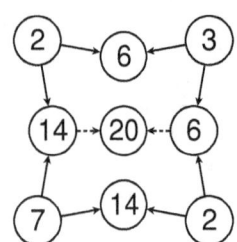

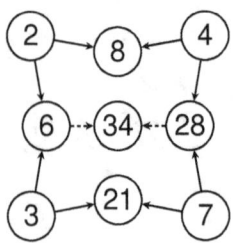 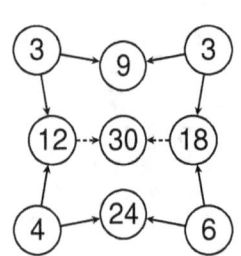

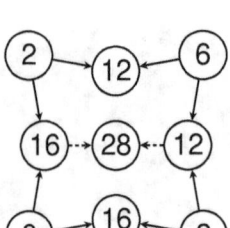 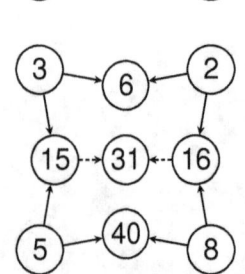

Solutions for Page 8

Find the missing numbers. Solid lines mean multiply. Dotted lines mean add.

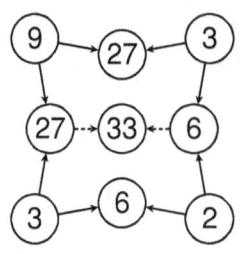 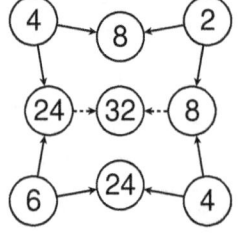

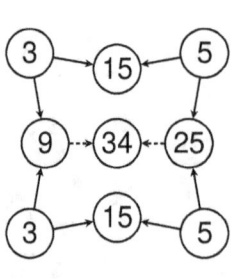 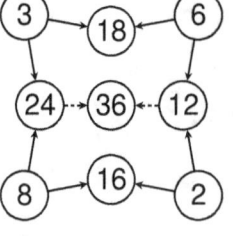

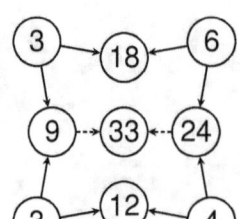 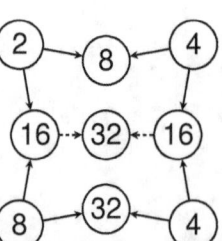

Solutions for Page 9

Find the missing numbers. Solid lines mean multiply.
Dotted lines mean add.

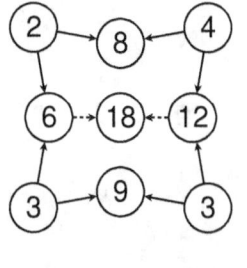 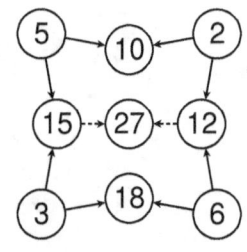

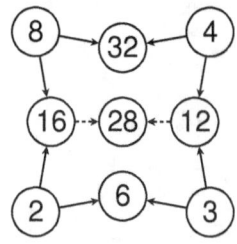 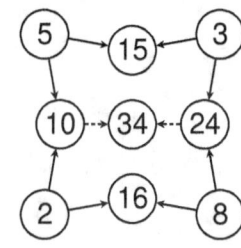

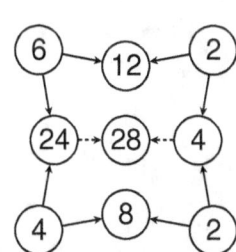 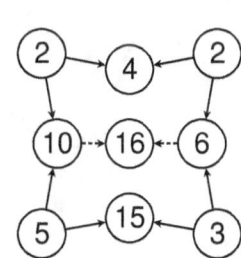

Solutions for Page 10

Find the missing numbers. Solid lines mean multiply.
Dotted lines mean add.

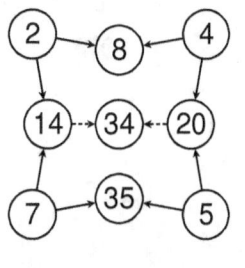 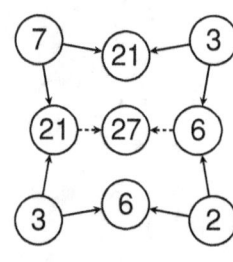

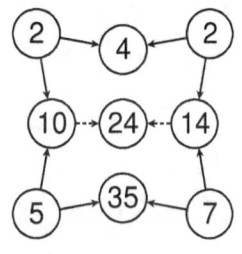 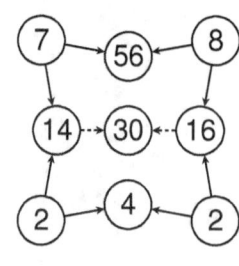

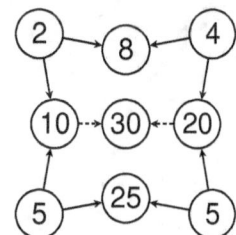 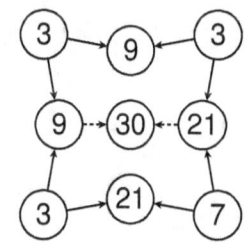

Solutions for Page 11

Find the missing numbers. Solid lines mean multiply.
Dotted lines mean add.

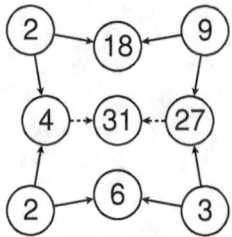 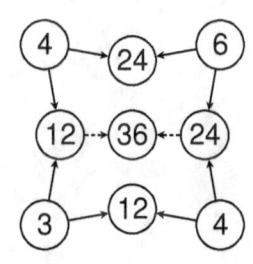

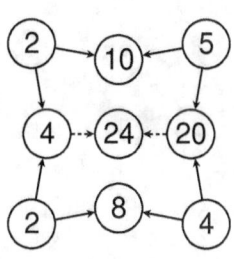 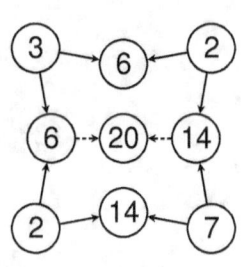

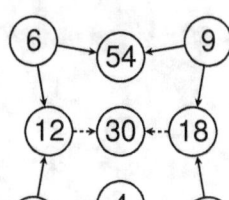 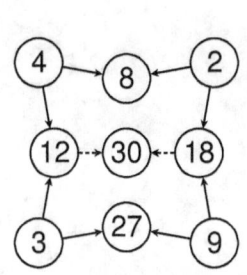

Solutions for Page 12

Find the missing numbers. Solid lines mean multiply.
Dotted lines mean add.

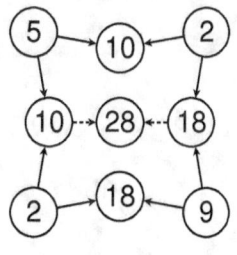 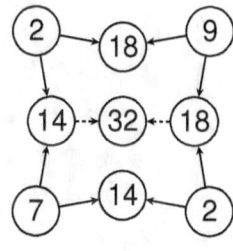

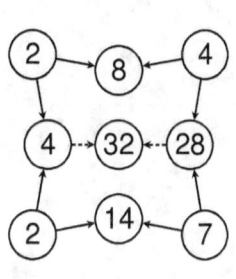 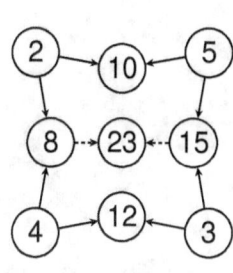

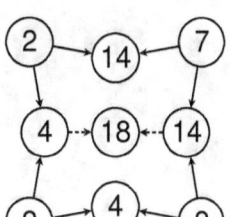 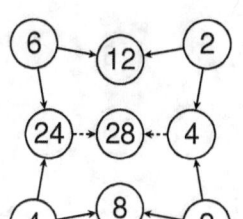

Solutions for Page 13

Find the missing numbers. Solid lines mean multiply.
Dotted lines mean add.

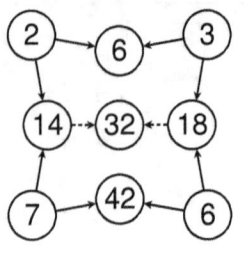 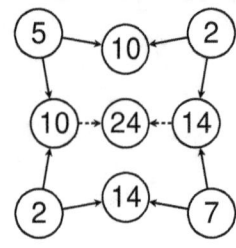

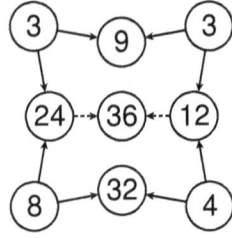 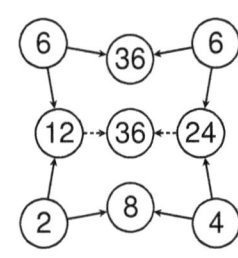

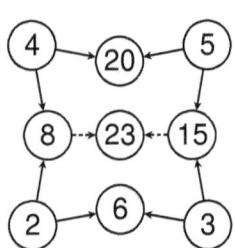 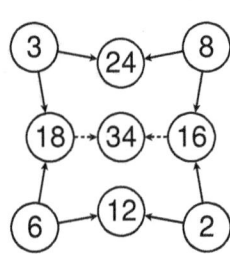

Solutions for Page 14

Find the missing numbers. Solid lines mean multiply.
Dotted lines mean add.

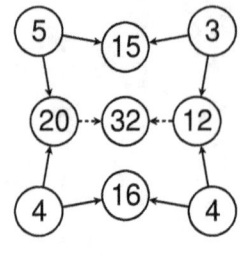

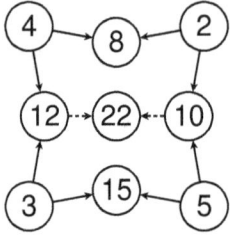 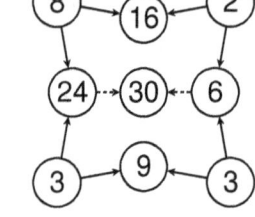

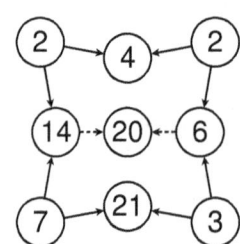 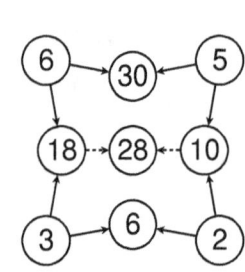

Solutions for Page 15

Find the missing numbers. Solid lines mean multiply.
Dotted lines mean add.

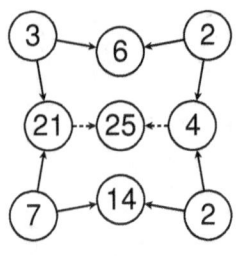 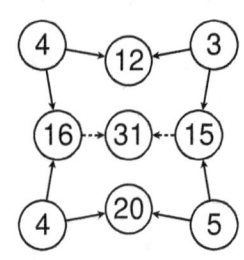

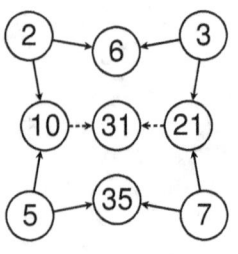 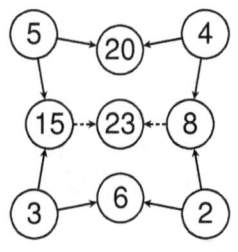

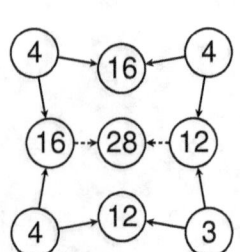 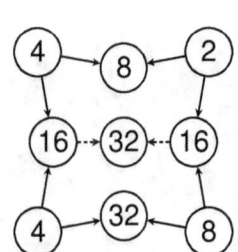

Solutions for Page 16

Find the missing numbers. Solid lines mean multiply.
Dotted lines mean add.

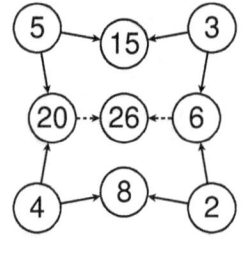 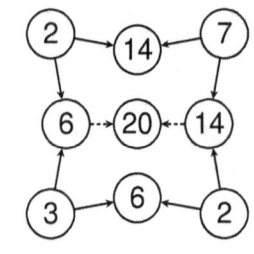

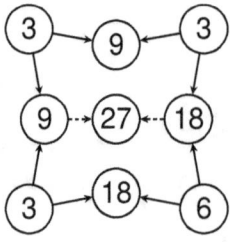 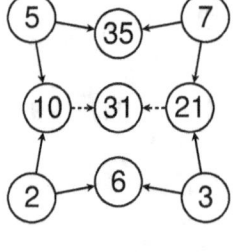

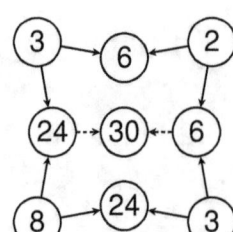 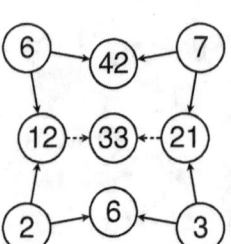

Solutions for Page 17

Find the missing numbers. Solid lines mean multiply.
Dotted lines mean add.

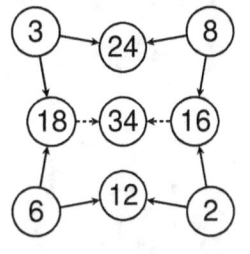 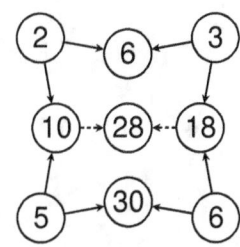

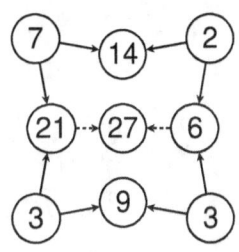 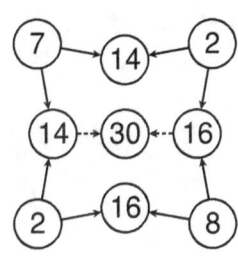

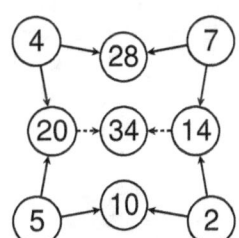 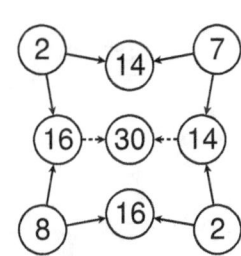

Solutions for Page 18

Find the missing numbers. Solid lines mean multiply.
Dotted lines mean add.

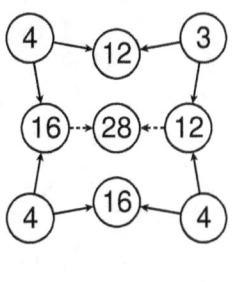 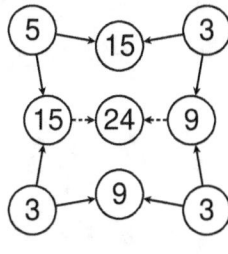

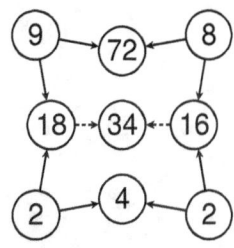 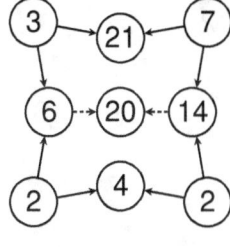

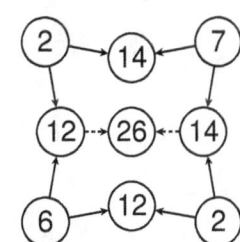 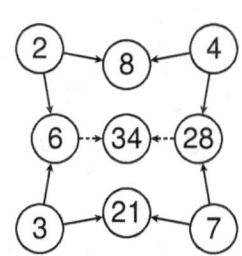

Solutions for Page 19

Find the missing numbers. Solid lines mean multiply.
Dotted lines mean add.

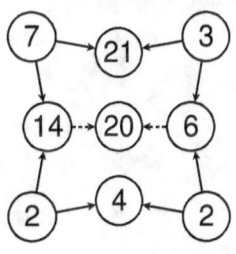 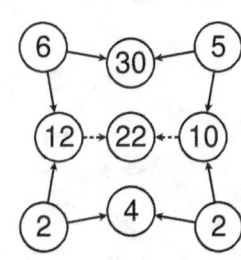

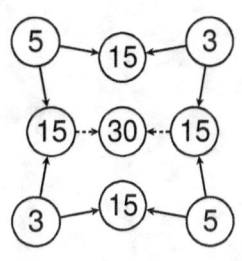 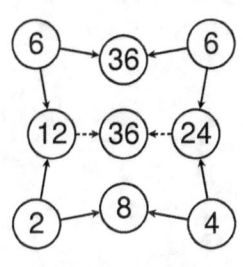

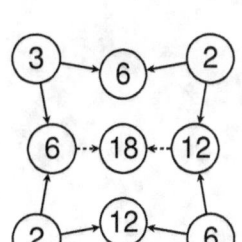 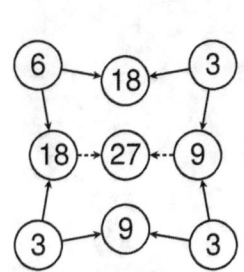

Solutions for Page 20

Find the missing numbers. Solid lines mean multiply.
Dotted lines mean add.

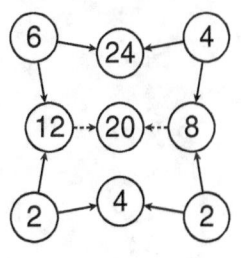 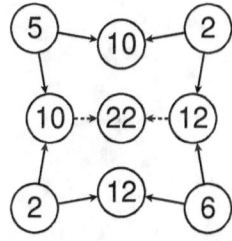

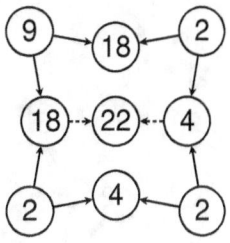 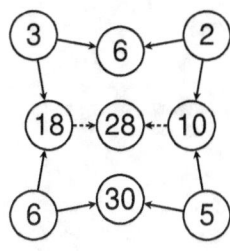

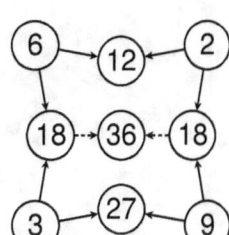 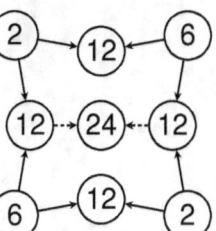

Solutions for Page 21

Find the missing numbers. Solid lines mean multiply.
Dotted lines mean add.

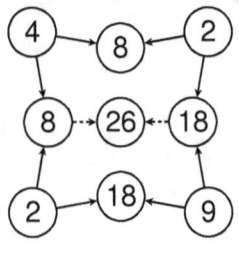 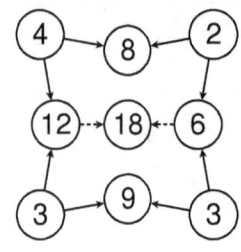

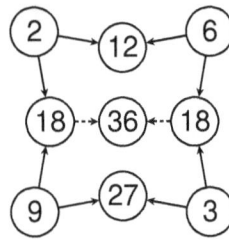 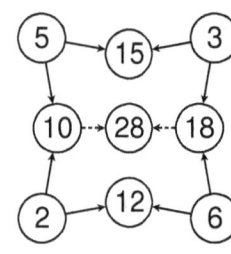

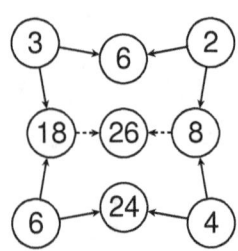 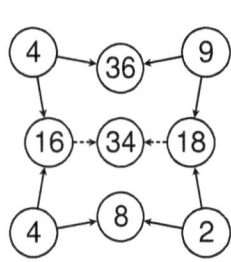

Solutions for Page 22

Find the missing numbers. Solid lines mean multiply.
Dotted lines mean add.

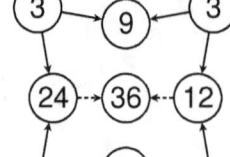 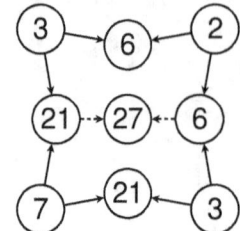

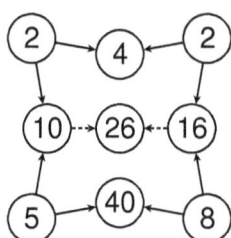 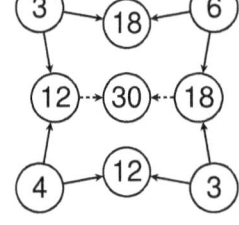

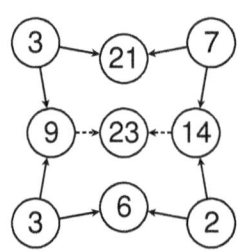 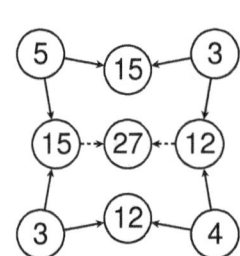

Solutions for Page 23

Find the missing numbers. Solid lines mean multiply.
Dotted lines mean add.

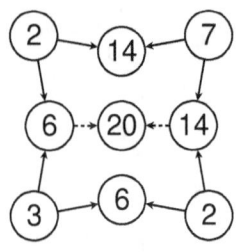 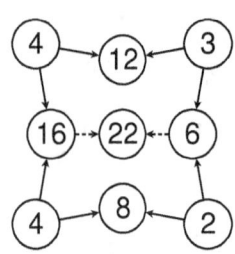

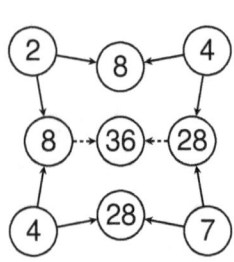 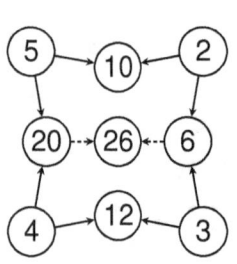

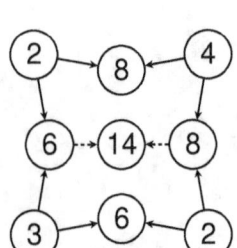 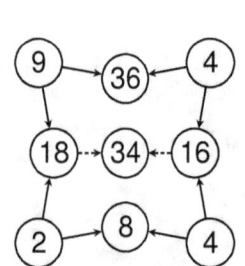

Solutions for Page 24

Find the missing numbers. Solid lines mean multiply.
Dotted lines mean add.

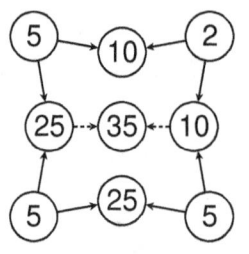 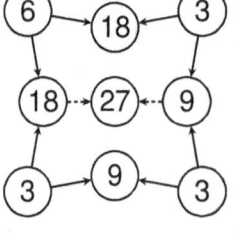

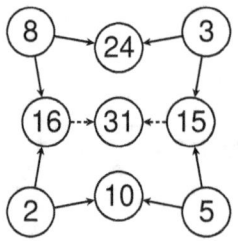 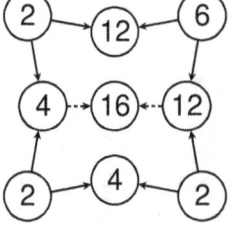

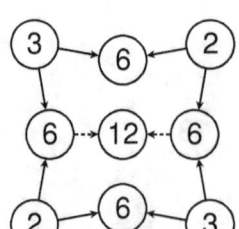 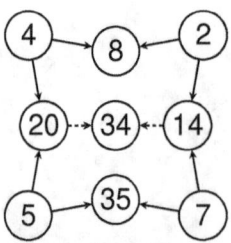

Solutions for Page 25

Find the missing numbers. Solid lines mean multiply.
Dotted lines mean add.

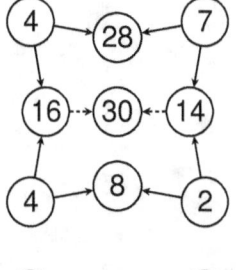 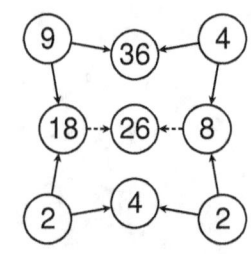

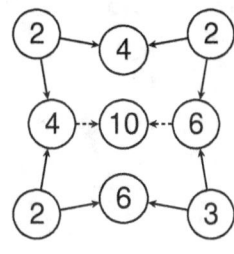 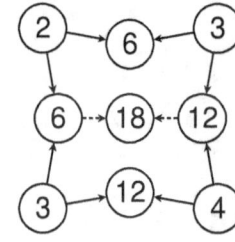

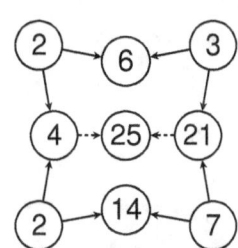 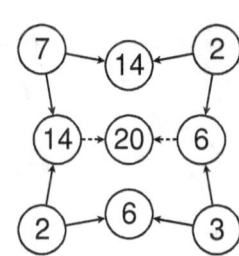

Solutions for Page 26

Find the missing numbers. Solid lines mean multiply.
Dotted lines mean add.

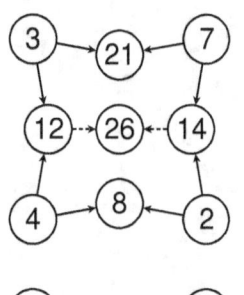 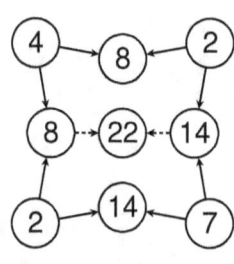

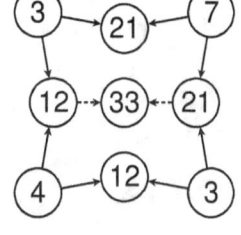 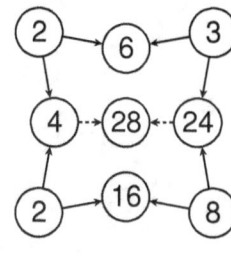

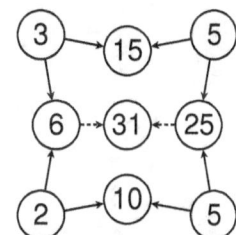 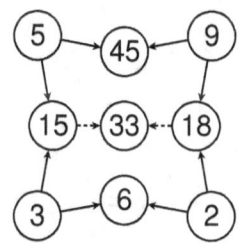

Solutions for Page 27

Find the missing numbers. Solid lines mean multiply.
Dotted lines mean add.

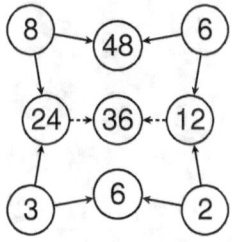 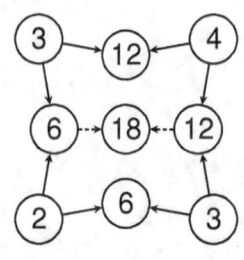

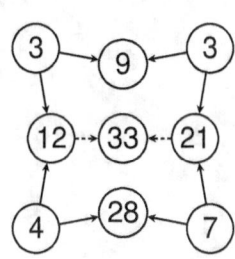 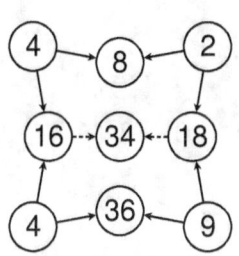

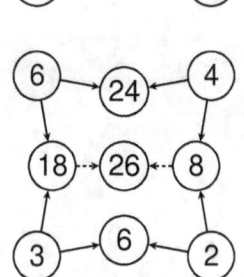 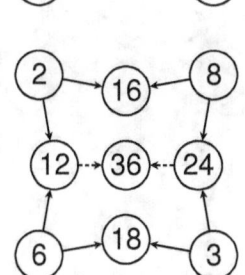

Solutions for Page 28

Find the missing numbers. Solid lines mean multiply.
Dotted lines mean add.

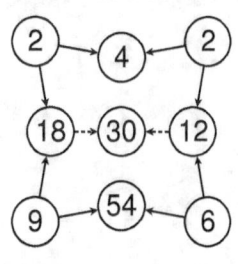 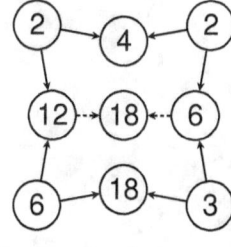

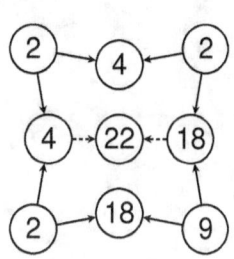 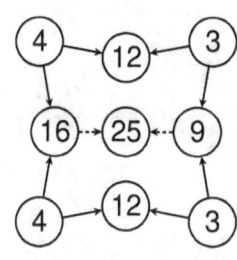

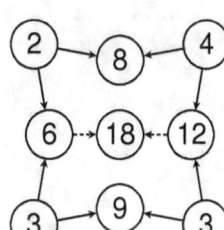 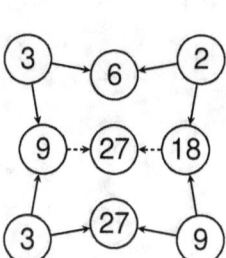

Solutions for Page 29

Find the missing numbers. Solid lines mean multiply.
Dotted lines mean add.

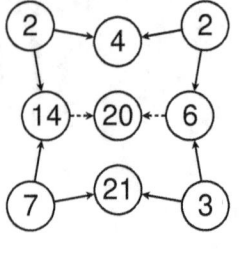

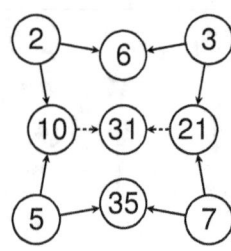

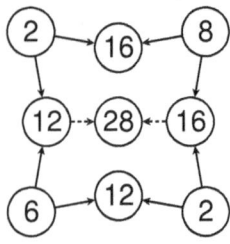

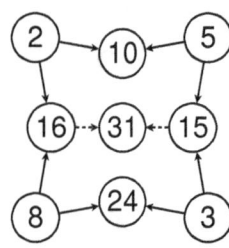

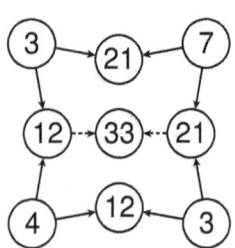

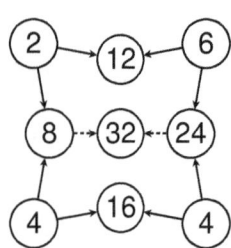

Solutions for Page 30

Find the missing numbers. Solid lines mean multiply.
Dotted lines mean add.

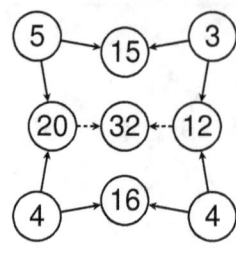

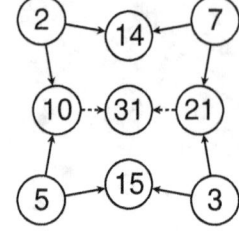

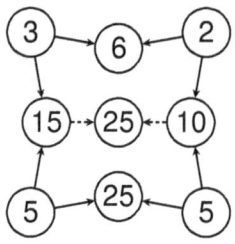

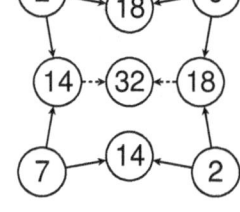

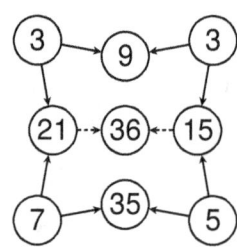

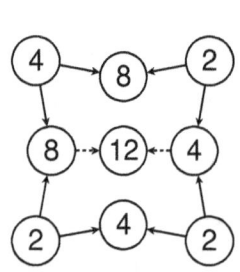

Solutions for Page 31

Find the missing numbers. Solid lines mean multiply.
Dotted lines mean add.

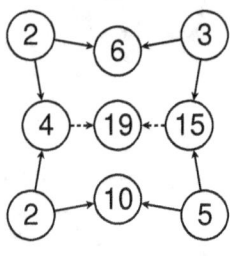

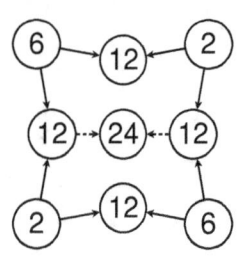

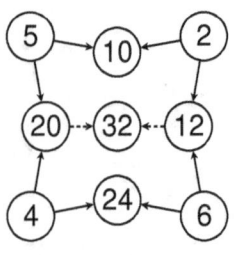

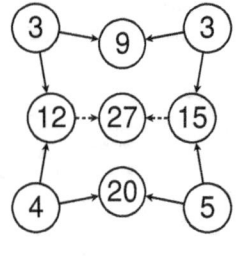

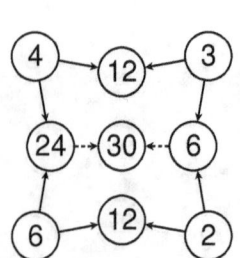

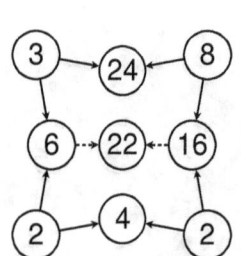

Solutions for Page 32

Find the missing numbers. Solid lines mean multiply.
Dotted lines mean add.

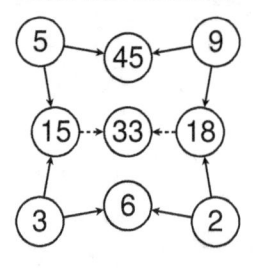

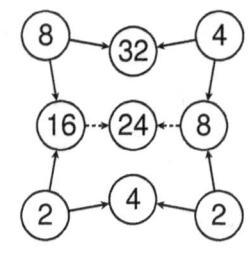

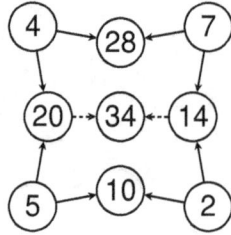

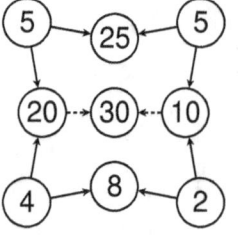

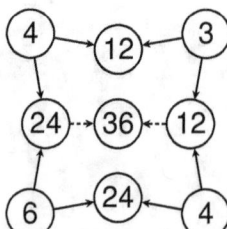

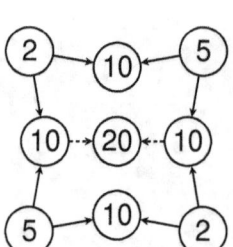

Solutions for Page 33

Find the missing numbers. Solid lines mean multiply.
Dotted lines mean add.

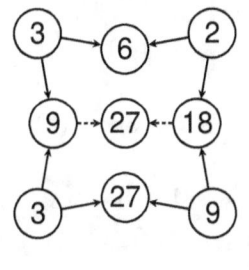
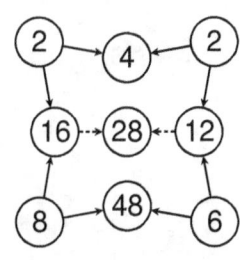
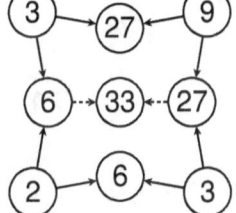
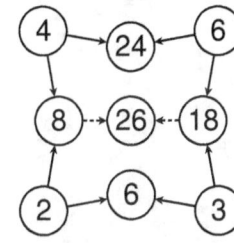
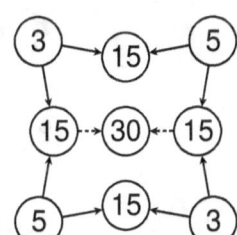
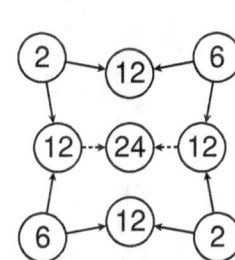

Solutions for Page 34

Find the missing numbers. Solid lines mean multiply.
Dotted lines mean add.

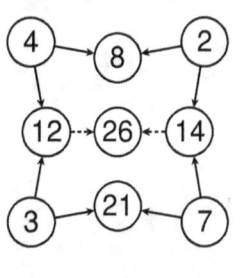
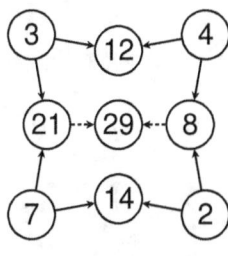
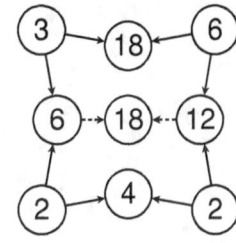
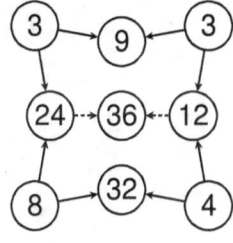
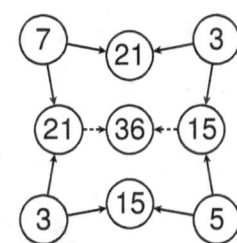
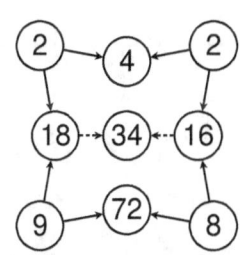

Solutions for Page 35

Find the missing numbers. Solid lines mean multiply.
Dotted lines mean add.

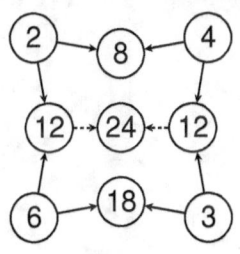
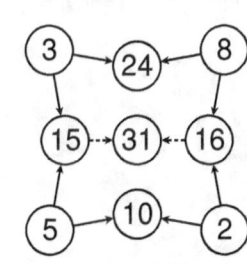
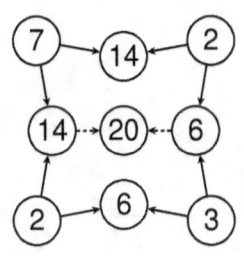
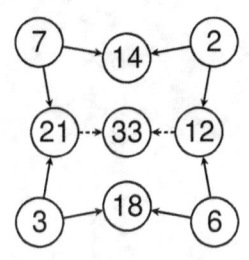
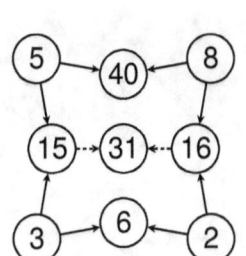
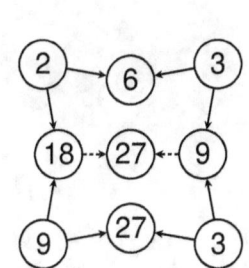

Solutions for Page 36

Find the missing numbers. Solid lines mean multiply.
Dotted lines mean add.

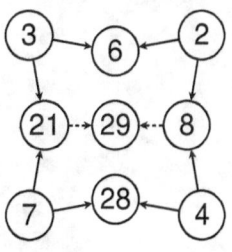
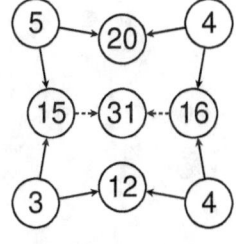
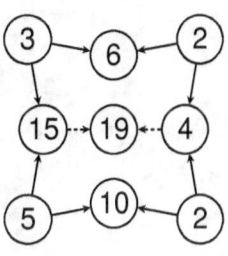
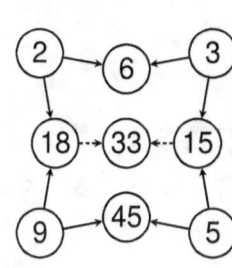
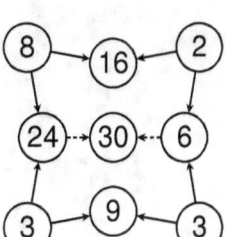
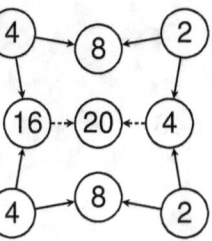

Solutions for Page 37

Find the missing numbers. Solid lines mean multiply.
Dotted lines mean add.

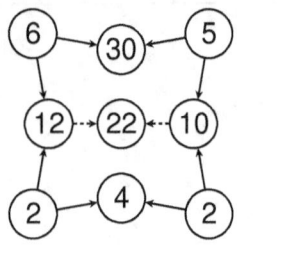 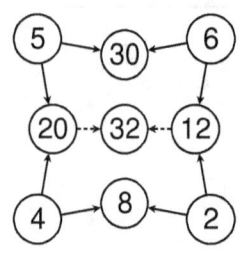

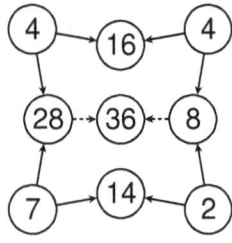 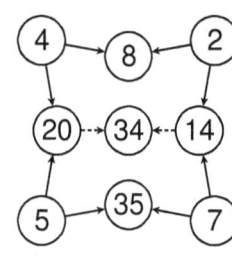

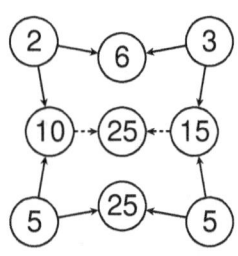 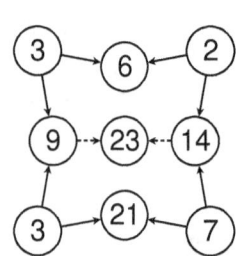

Solutions for Page 38

Find the missing numbers. Solid lines mean multiply.
Dotted lines mean add.

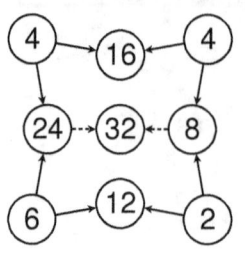 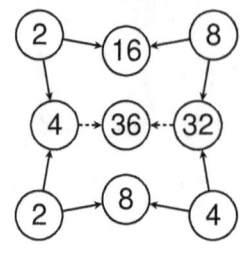

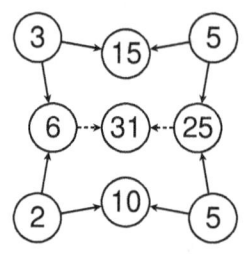 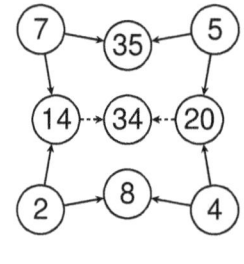

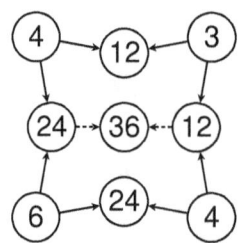 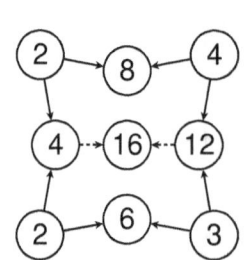

Solutions for Page 39

Find the missing numbers. Solid lines mean multiply.
Dotted lines mean add.

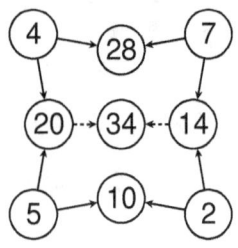 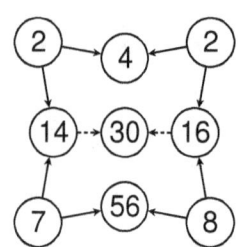

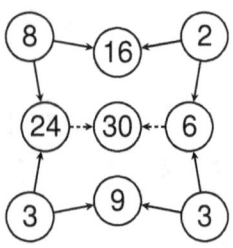 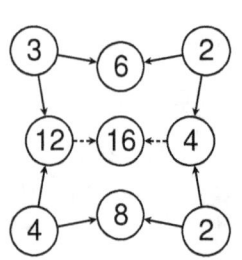

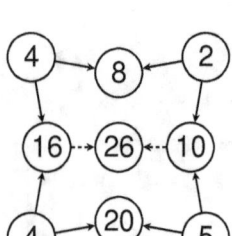 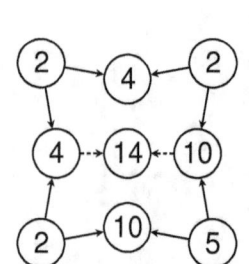

Solutions for Page 40

Find the missing numbers. Solid lines mean multiply.
Dotted lines mean add.

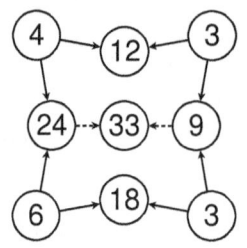 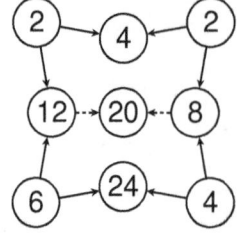

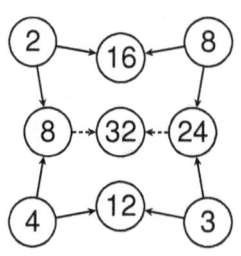 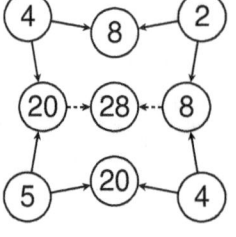

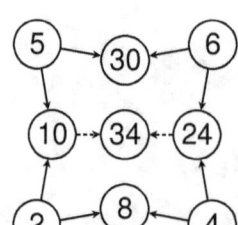 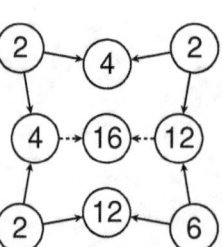

Solutions for Page 41

Find the missing numbers. Solid lines mean multiply.
Dotted lines mean add.

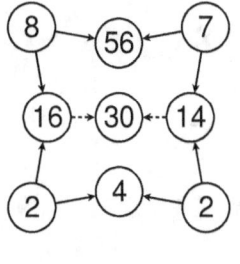
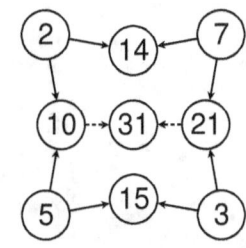
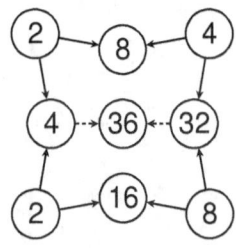
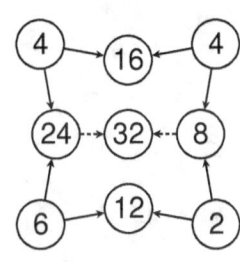
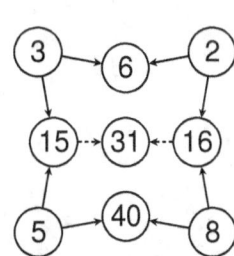
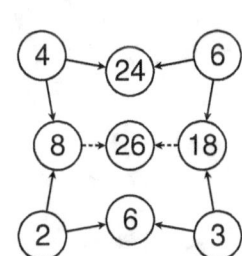

Solutions for Page 42

Find the missing numbers. Solid lines mean multiply.
Dotted lines mean add.

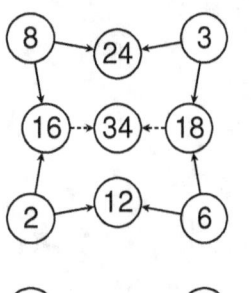
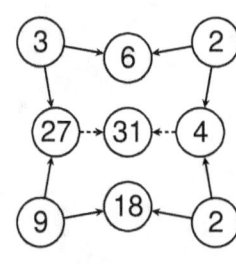
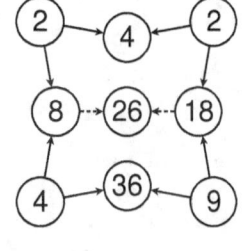
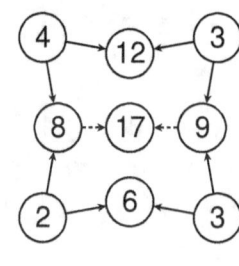
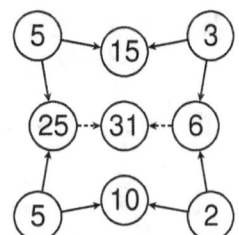
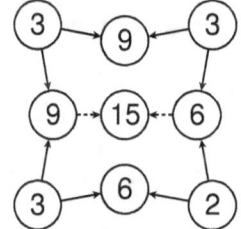

Solutions for Page 43

Find the missing numbers. Solid lines mean multiply.
Dotted lines mean add.

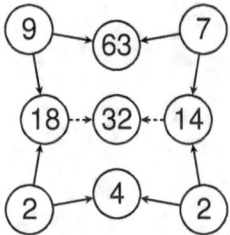
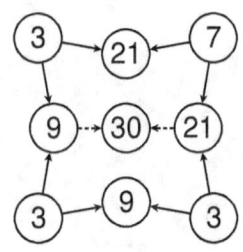
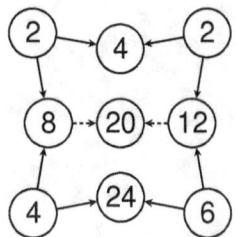
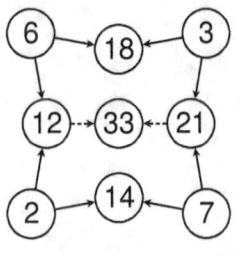
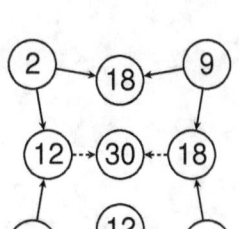
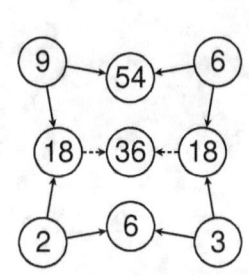

Solutions for Page 44

Find the missing numbers. Solid lines mean multiply.
Dotted lines mean add.

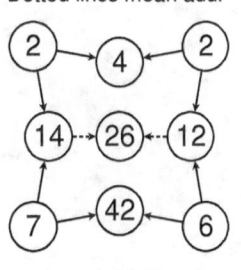
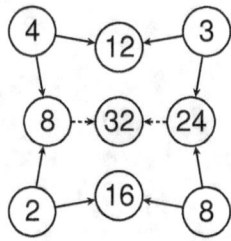
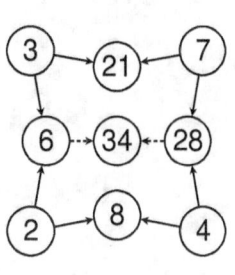
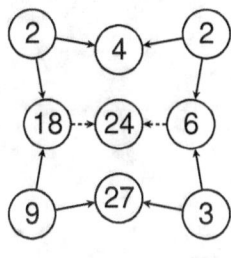
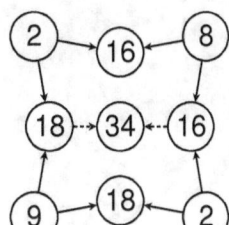
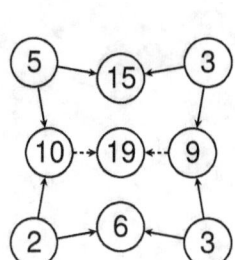

Solutions for Page 45

Find the missing numbers. Solid lines mean multiply.
Dotted lines mean add.

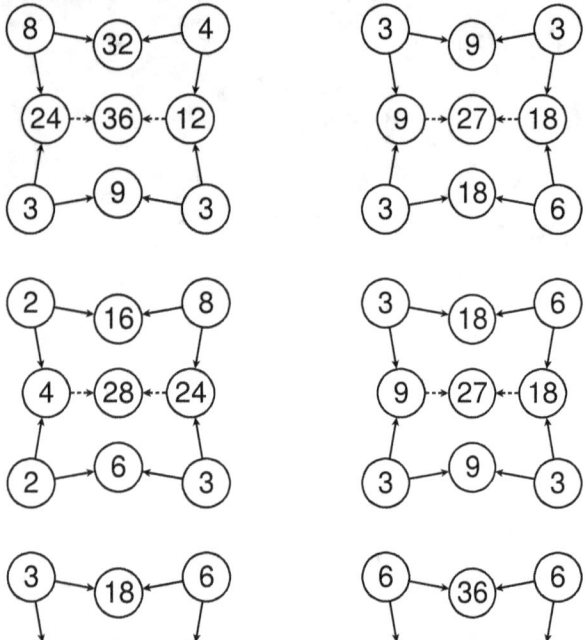

Solutions for Page 46

Find the missing numbers. Solid lines mean multiply.
Dotted lines mean add.

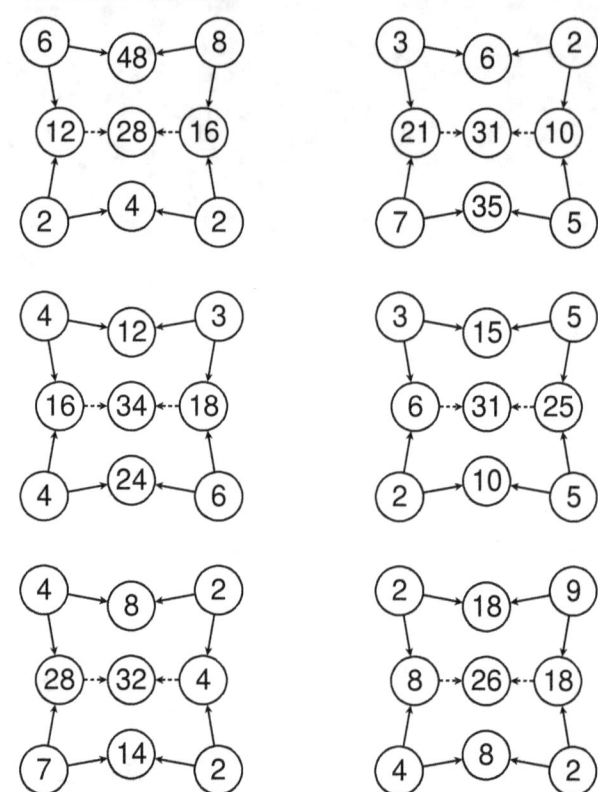

Solutions for Page 47

Find the missing numbers. Solid lines mean multiply.
Dotted lines mean add.

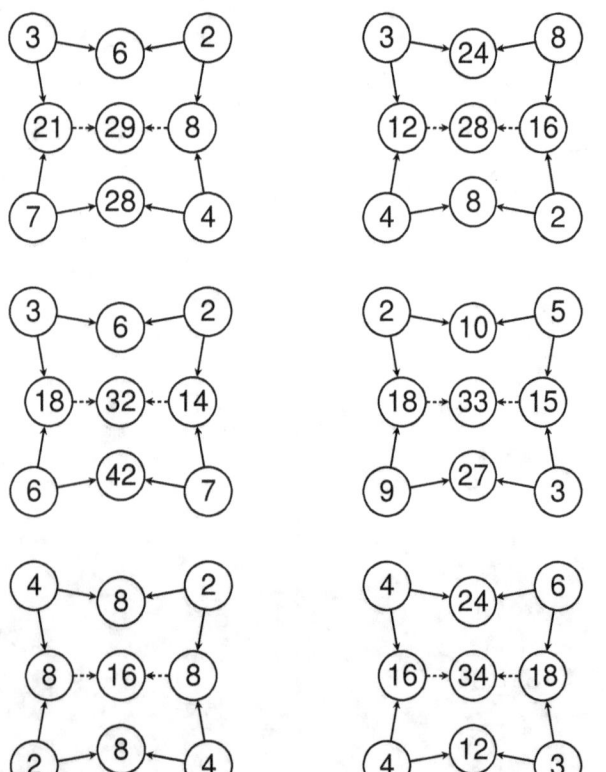

Solutions for Page 48

Find the missing numbers. Solid lines mean multiply.
Dotted lines mean add.

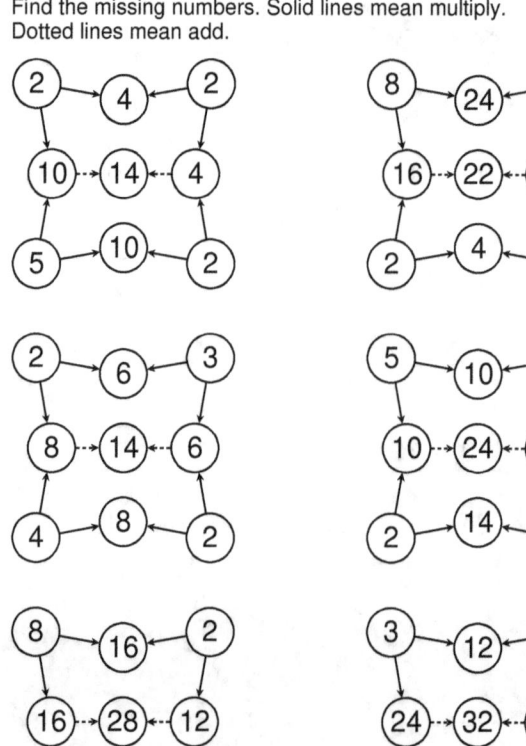

Solutions for Page 49

Find the missing numbers. Solid lines mean multiply.
Dotted lines mean add.

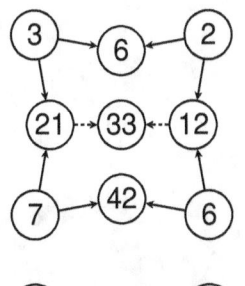 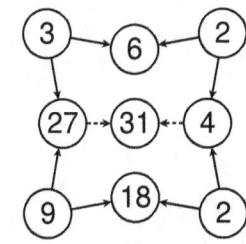

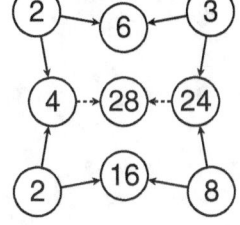 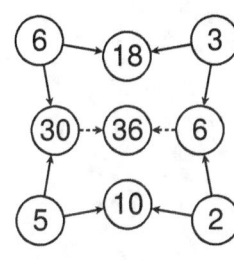

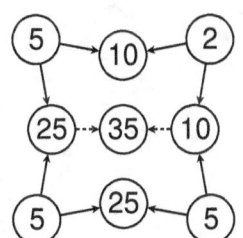 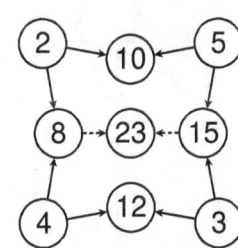

Solutions for Page 50

Find the missing numbers. Solid lines mean multiply.
Dotted lines mean add.

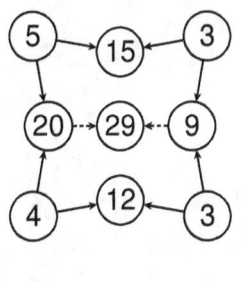 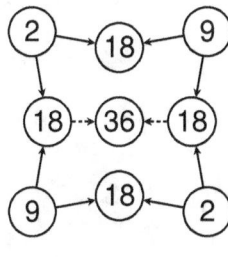

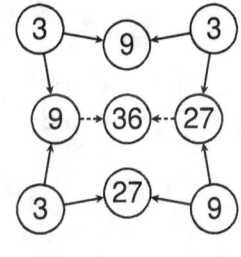 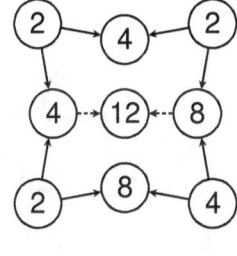

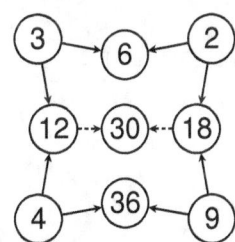 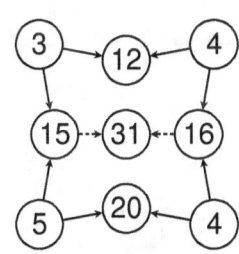

Solutions for Page 51

Find the missing numbers. Solid lines mean multiply.
Dotted lines mean add.

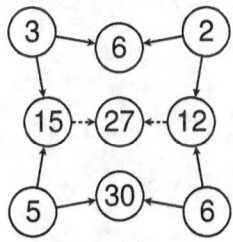 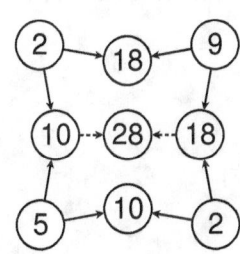

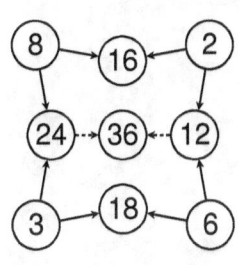 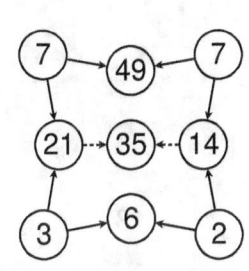

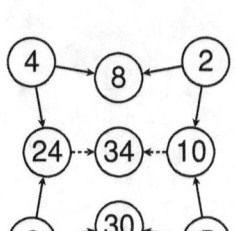 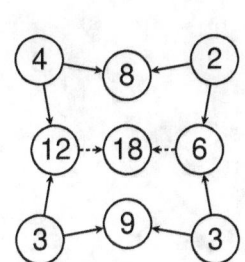

Solutions for Page 52

Find the missing numbers. Solid lines mean multiply.
Dotted lines mean add.

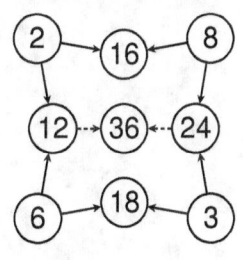 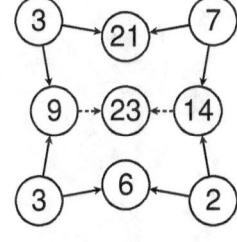

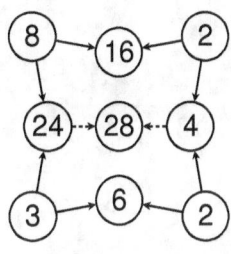 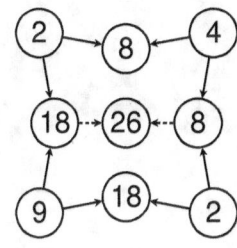

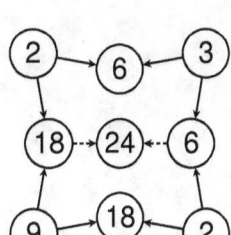 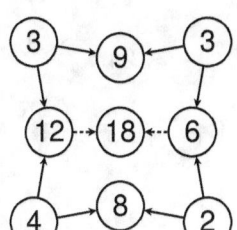

Solutions for Page 53

Find the missing numbers. Solid lines mean multiply.
Dotted lines mean add.

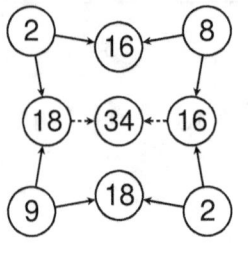 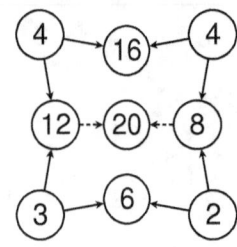

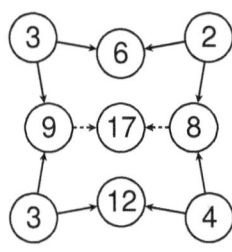 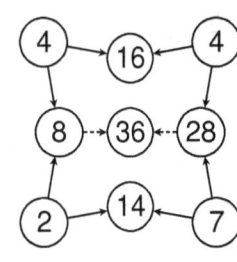

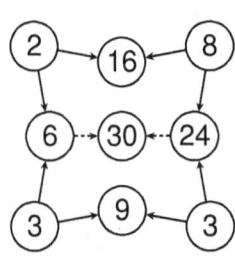 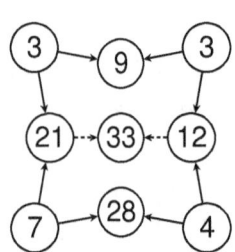

Solutions for Page 54

Find the missing numbers. Solid lines mean multiply.
Dotted lines mean add.

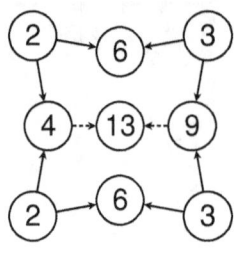 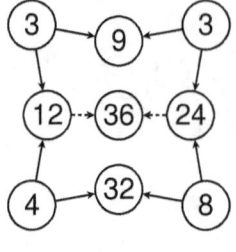

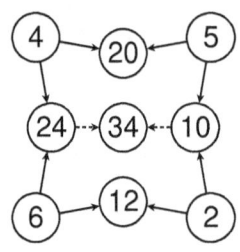 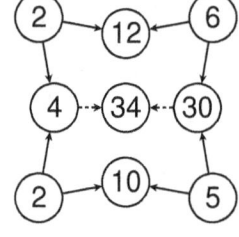

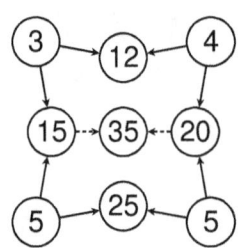 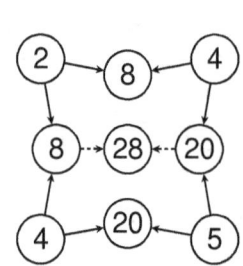

Solutions for Page 55

Find the missing numbers. Solid lines mean multiply.
Dotted lines mean add.

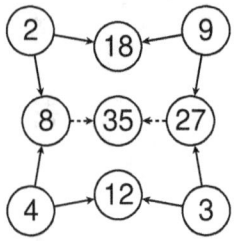 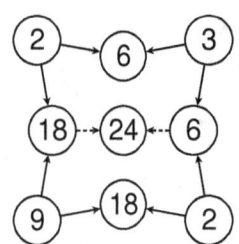

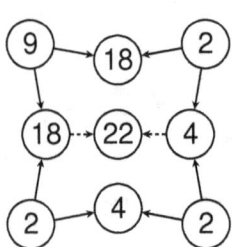 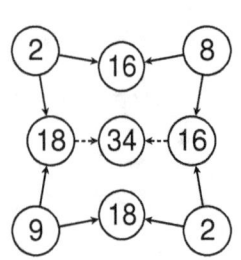

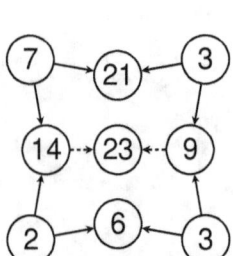 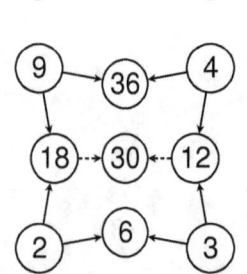

Solutions for Page 56

Find the missing numbers. Solid lines mean multiply.
Dotted lines mean add.

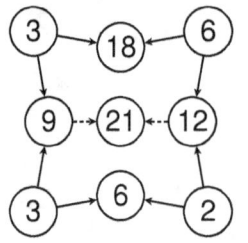 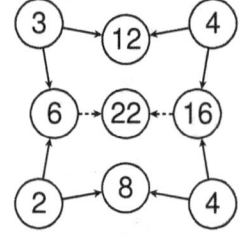

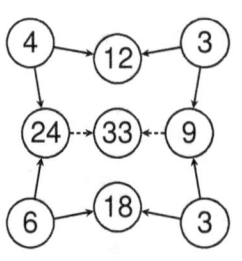 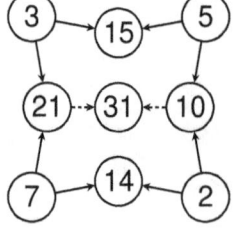

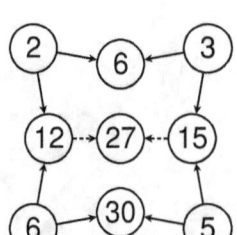 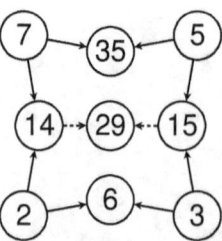

Solutions for Page 57

Find the missing numbers. Solid lines mean multiply.
Dotted lines mean add.

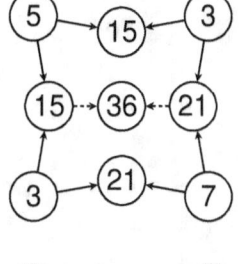
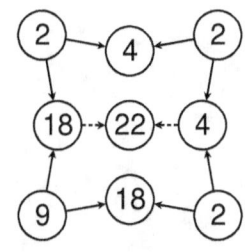
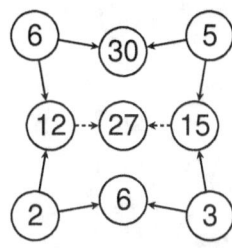
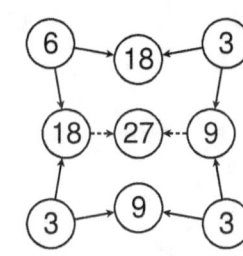
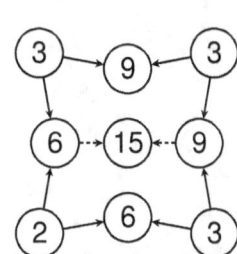
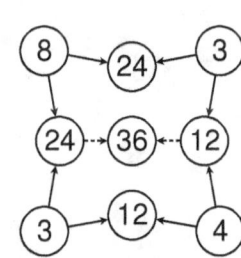

Solutions for Page 58

Find the missing numbers. Solid lines mean multiply.
Dotted lines mean add.

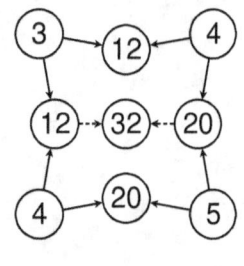
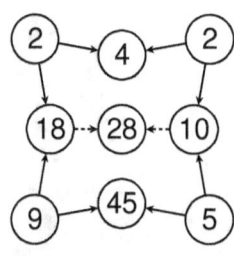
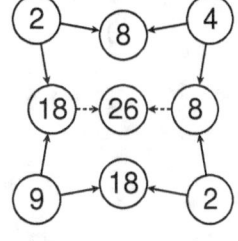
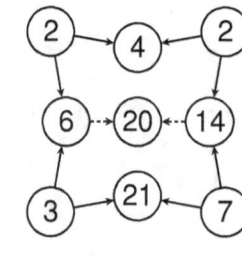
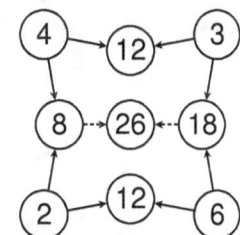
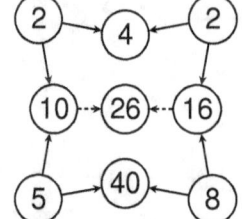

Solutions for Page 59

Find the missing numbers. Solid lines mean multiply.
Dotted lines mean add.

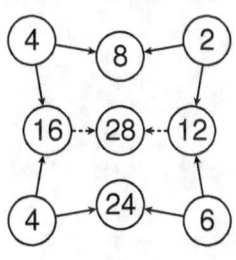
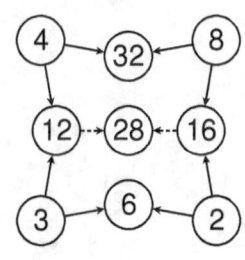
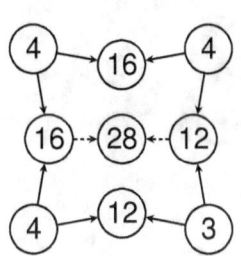
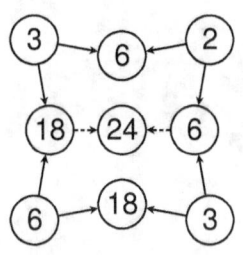
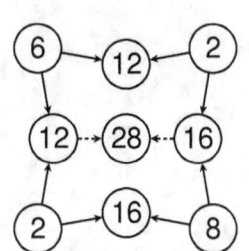
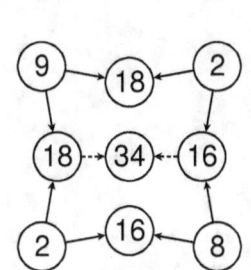

Solutions for Page 60

Find the missing numbers. Solid lines mean multiply.
Dotted lines mean add.

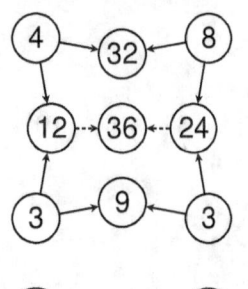
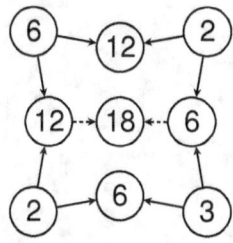
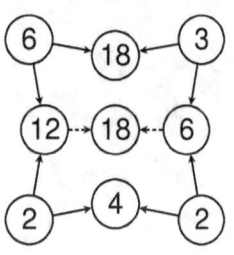
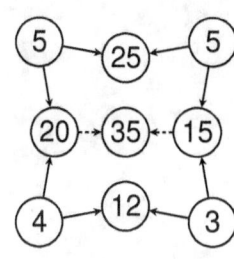
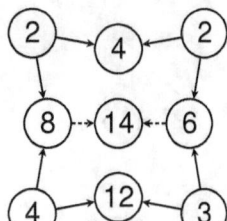
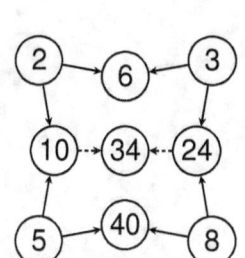

Solutions for Page 61

Find the missing numbers. Solid lines mean multiply.
Dotted lines mean add.

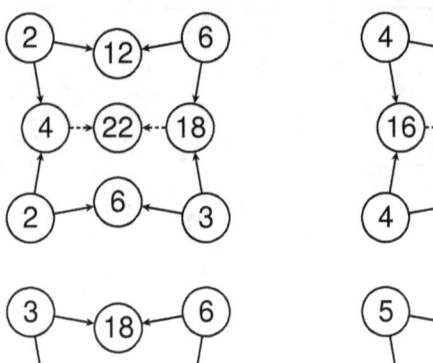
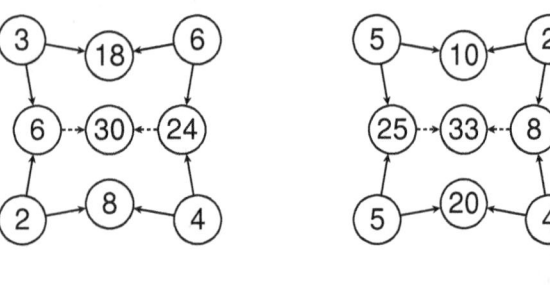
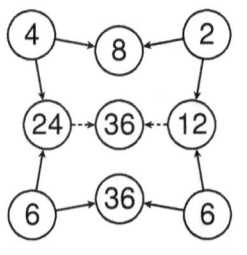
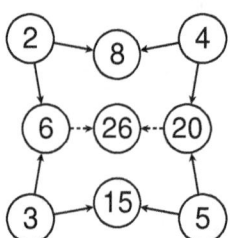
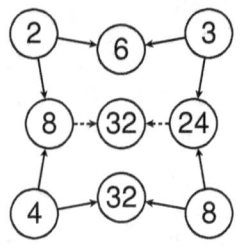
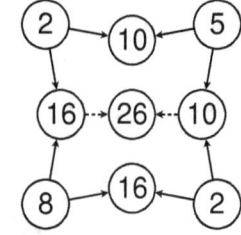
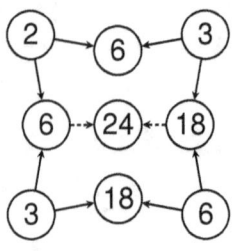
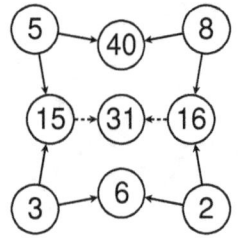
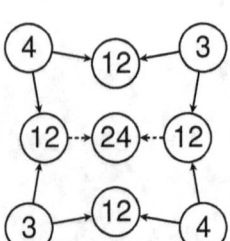
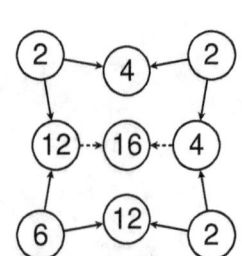

Solutions for Page 62

Find the missing numbers. Solid lines mean multiply.
Dotted lines mean add.

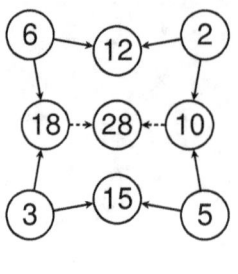
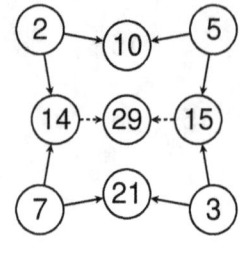
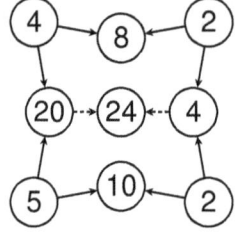
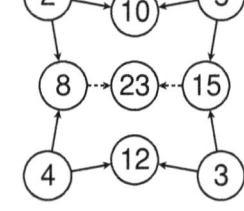
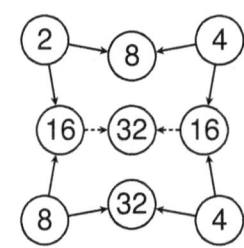
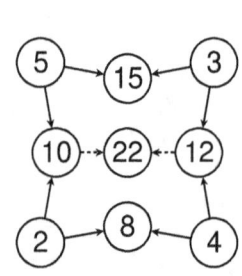
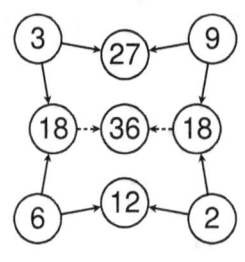
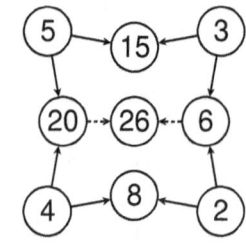
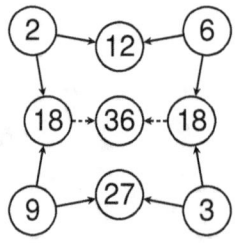
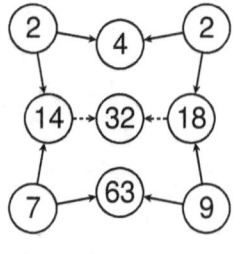
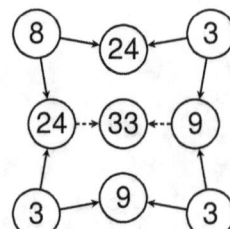
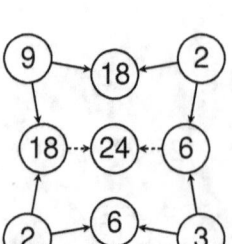

Solutions for Page 63

Find the missing numbers. Solid lines mean multiply.
Dotted lines mean add.

Solutions for Page 64

Find the missing numbers. Solid lines mean multiply.
Dotted lines mean add.

Solutions for Page 65

Find the missing numbers. Solid lines mean multiply.
Dotted lines mean add.

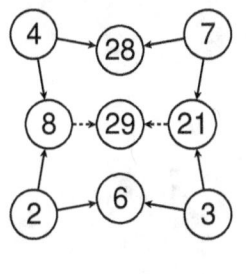
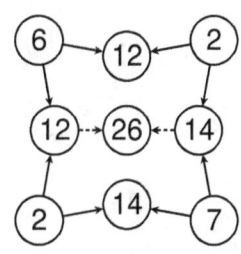
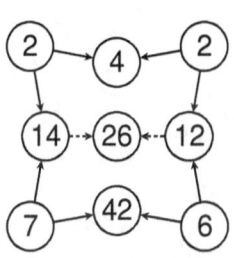
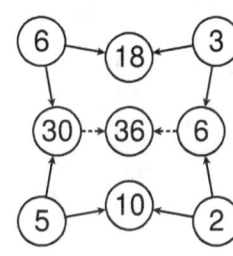
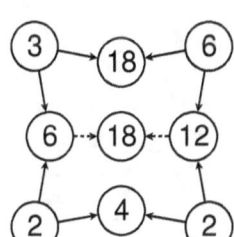
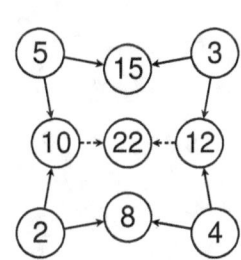

Solutions for Page 66

Find the missing numbers. Solid lines mean multiply.
Dotted lines mean add.

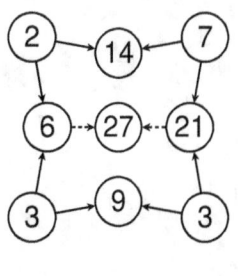
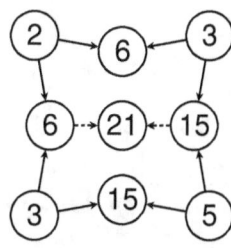
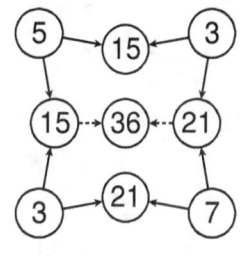
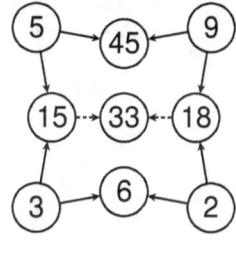
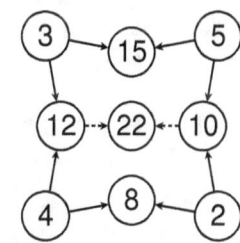
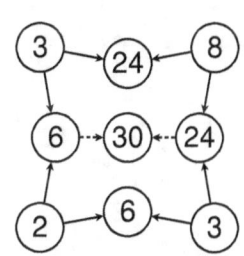

Solutions for Page 67

Find the missing numbers. Solid lines mean multiply.
Dotted lines mean add.

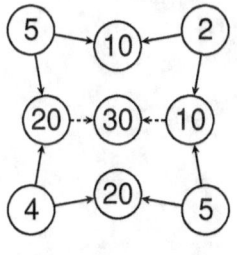
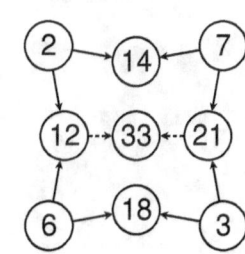
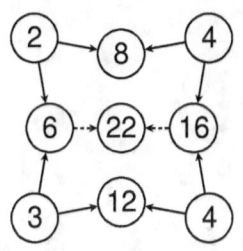
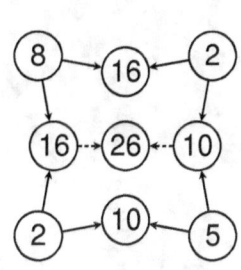
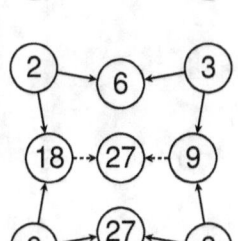
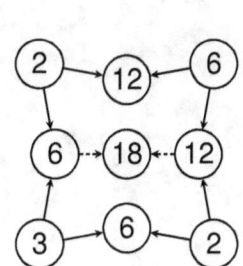

Solutions for Page 68

Find the missing numbers. Solid lines mean multiply.
Dotted lines mean add.

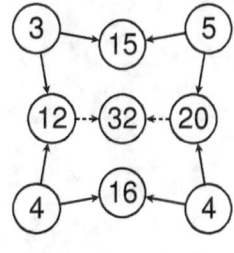
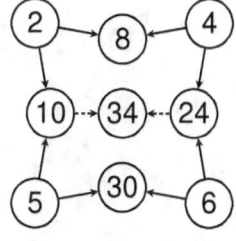
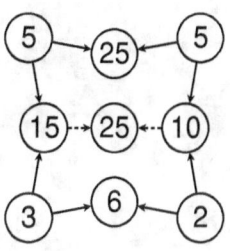
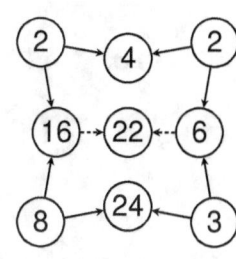
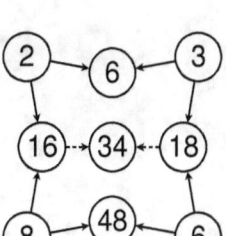
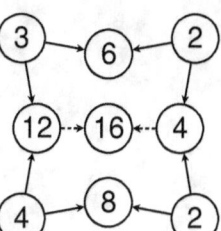

Solutions for Page 69

Find the missing numbers. Solid lines mean multiply.
Dotted lines mean add.

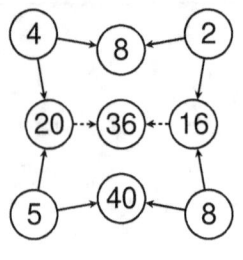 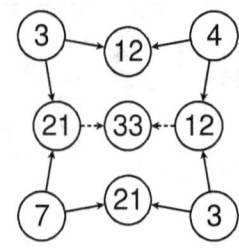

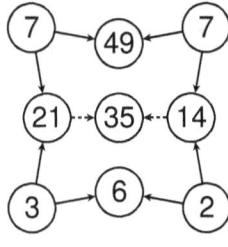 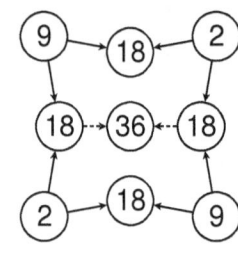

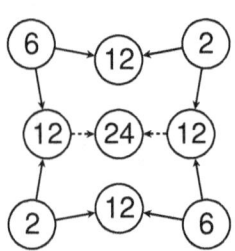 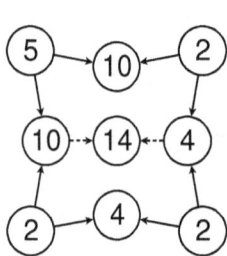

Solutions for Page 70

Find the missing numbers. Solid lines mean multiply.
Dotted lines mean add.

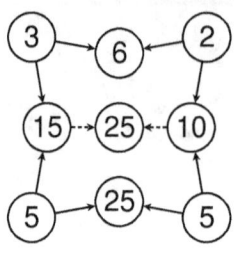 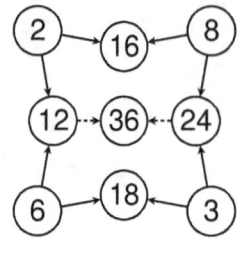

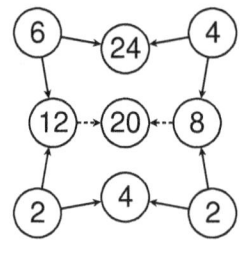 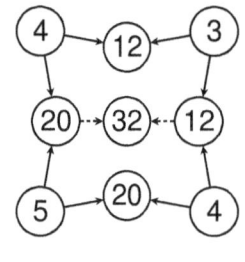

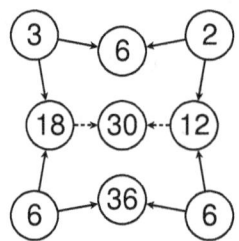 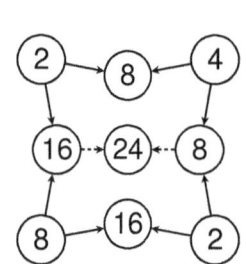

Solutions for Page 71

Find the missing numbers. Solid lines mean multiply.
Dotted lines mean add.

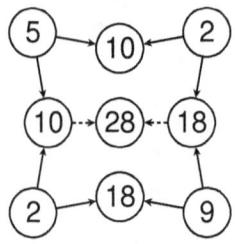 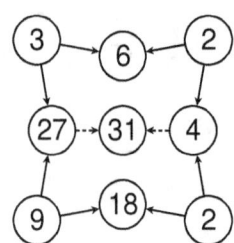

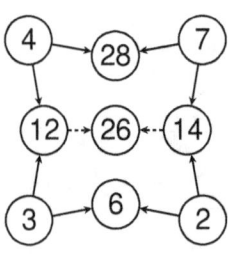 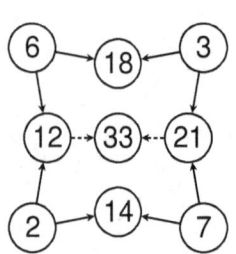

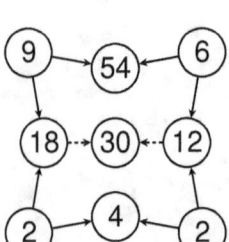 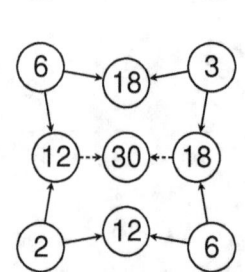

Solutions for Page 72

Find the missing numbers. Solid lines mean multiply.
Dotted lines mean add.

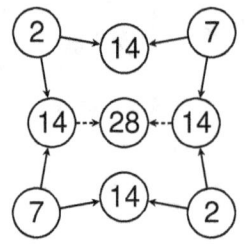 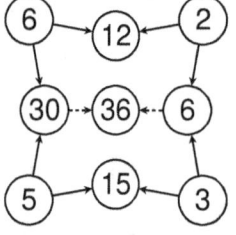

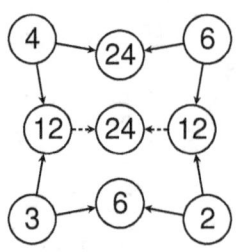 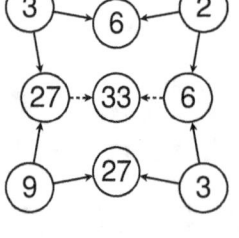

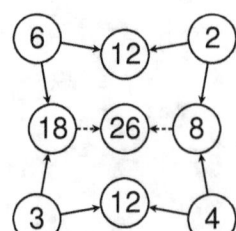 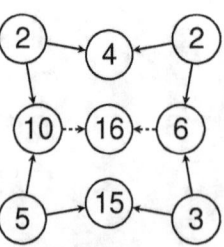

Solutions for Page 73

Find the missing numbers. Solid lines mean multiply.
Dotted lines mean add.

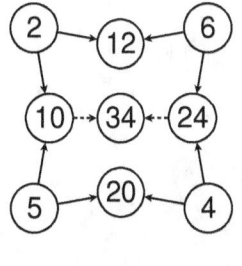
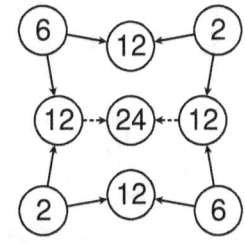

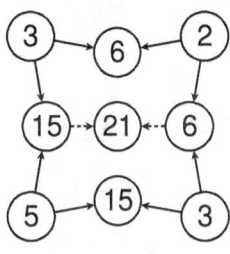
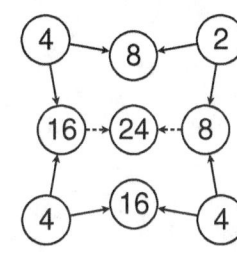

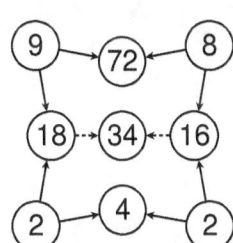
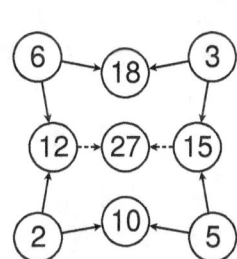

Solutions for Page 74

Find the missing numbers. Solid lines mean multiply.
Dotted lines mean add.

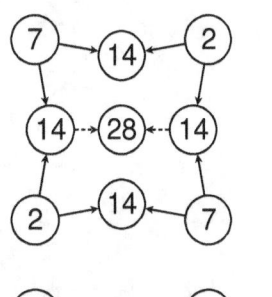
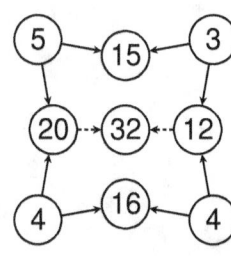

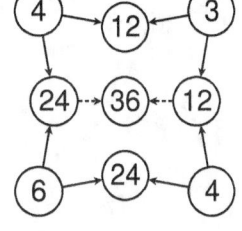
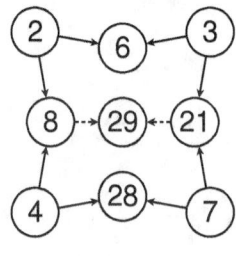

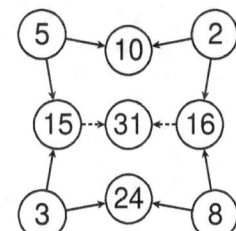
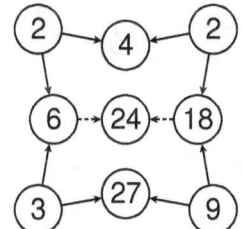

Solutions for Page 75

Find the missing numbers. Solid lines mean multiply.
Dotted lines mean add.

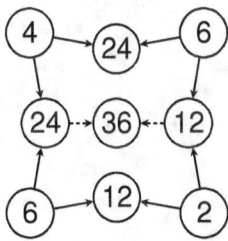
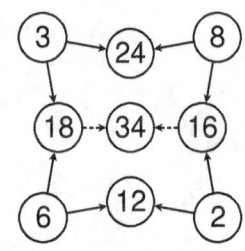

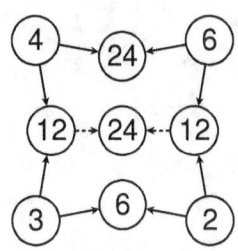
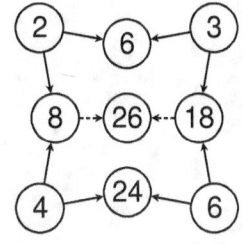

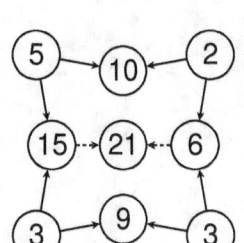
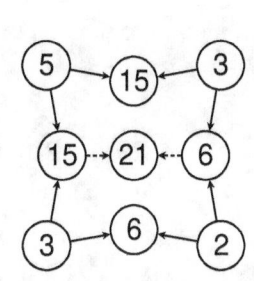

Solutions for Page 76

Find the missing numbers. Solid lines mean multiply.
Dotted lines mean add.

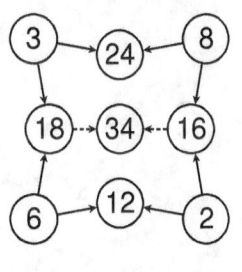
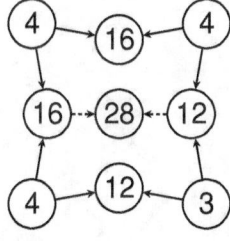

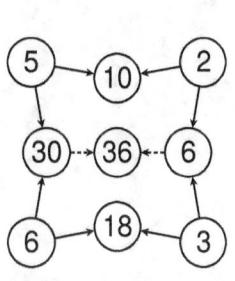
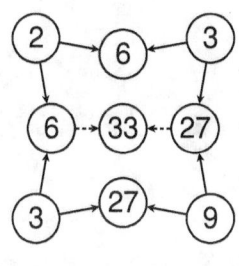

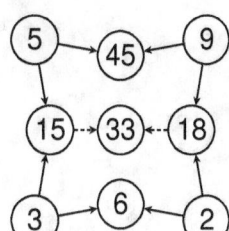
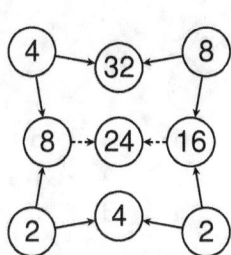

Solutions for Page 77

Find the missing numbers. Solid lines mean multiply.
Dotted lines mean add.

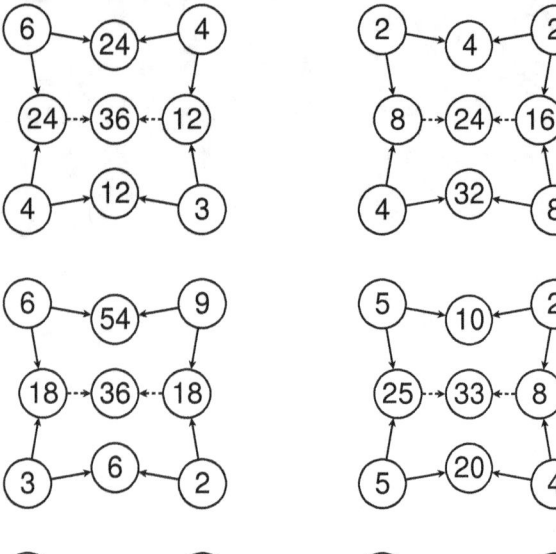

Solutions for Page 78

Find the missing numbers. Solid lines mean multiply.
Dotted lines mean add.

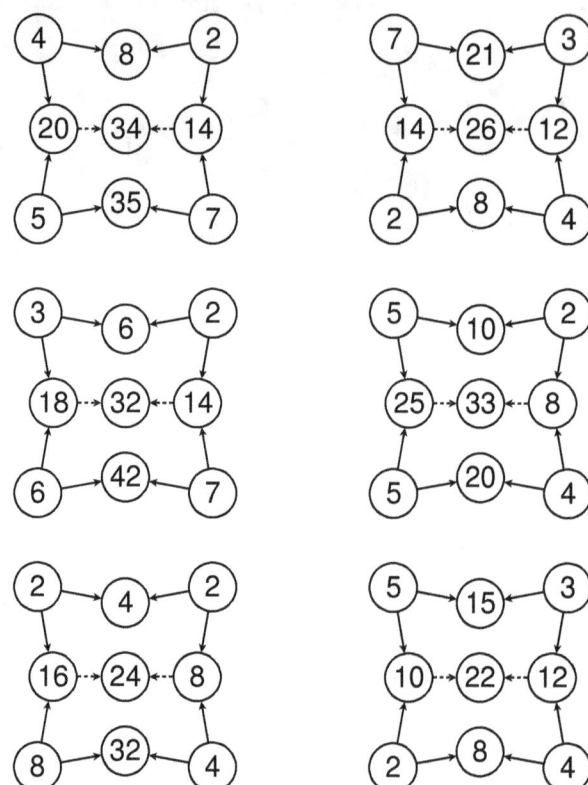

Solutions for Page 79

Find the missing numbers. Solid lines mean multiply.
Dotted lines mean add.

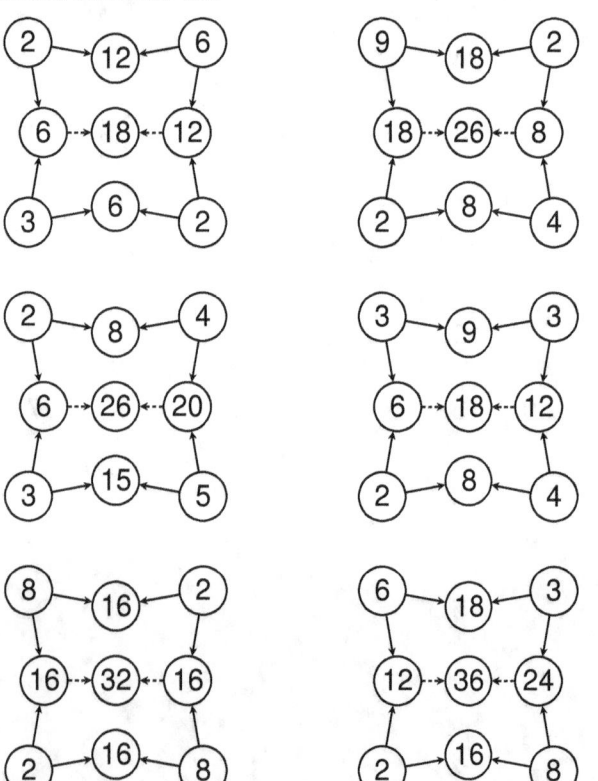

Solutions for Page 80

Find the missing numbers. Solid lines mean multiply.
Dotted lines mean add.

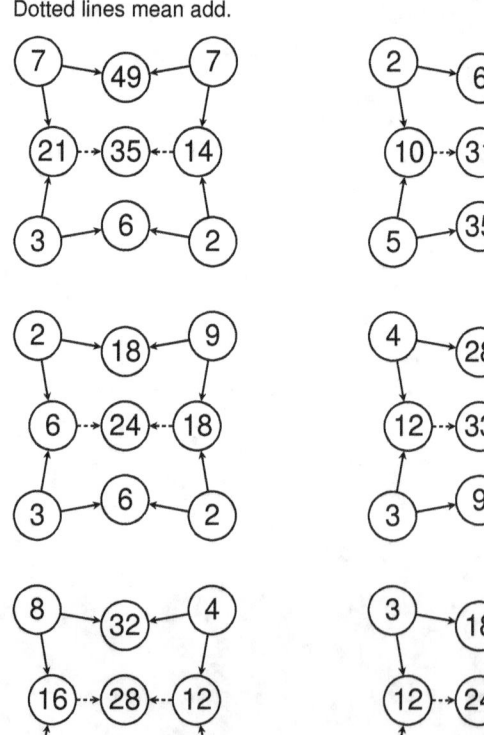

Solutions for Page 81

Find the missing numbers. Solid lines mean multiply.
Dotted lines mean add.

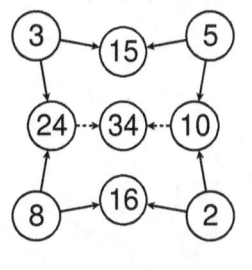
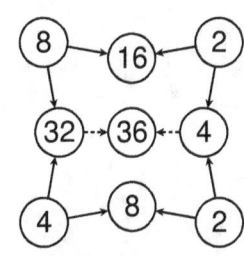

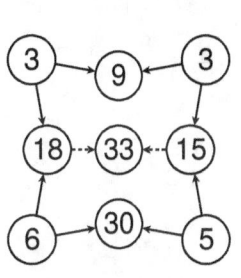
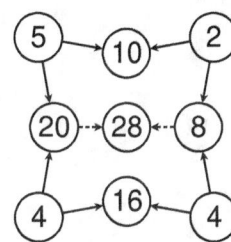

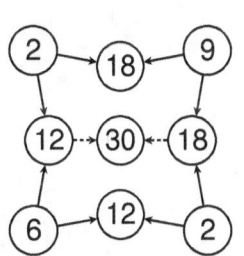
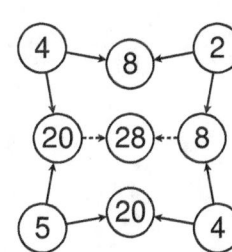

Solutions for Page 82

Find the missing numbers. Solid lines mean multiply.
Dotted lines mean add.

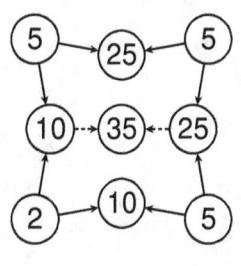
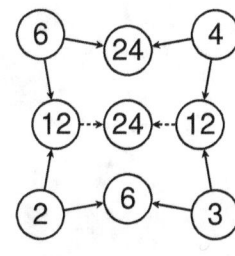

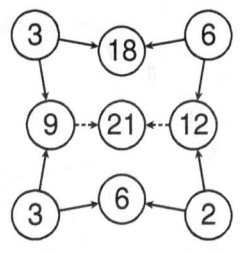
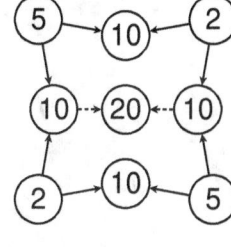

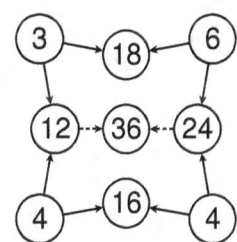
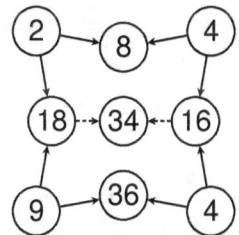

Solutions for Page 83

Find the missing numbers. Solid lines mean multiply.
Dotted lines mean add.

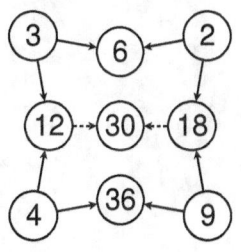
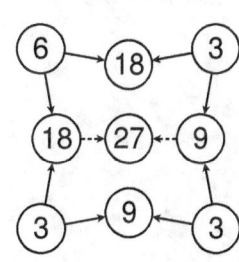

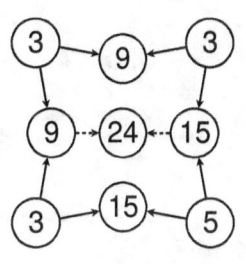
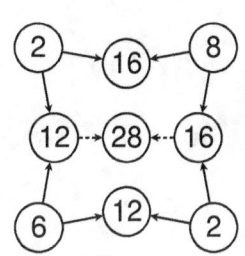

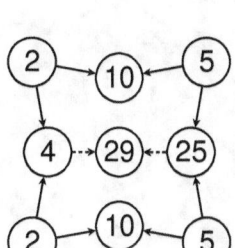
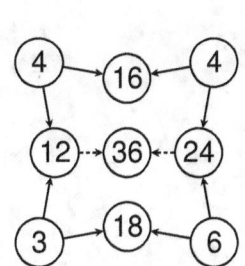

Solutions for Page 84

Find the missing numbers. Solid lines mean multiply.
Dotted lines mean add.

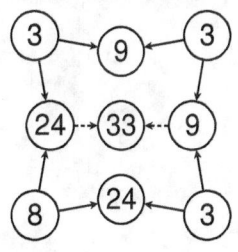
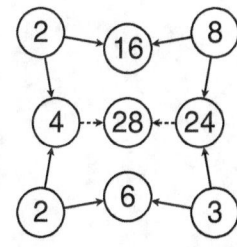

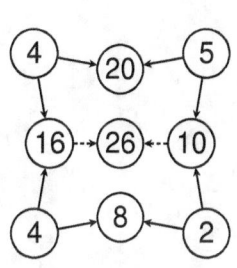
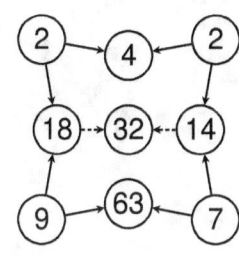

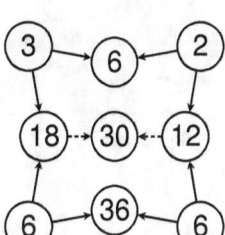
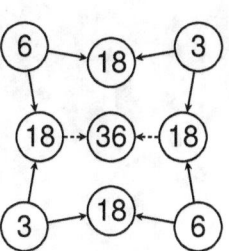

Solutions for Page 85

Find the missing numbers. Solid lines mean multiply.
Dotted lines mean add.

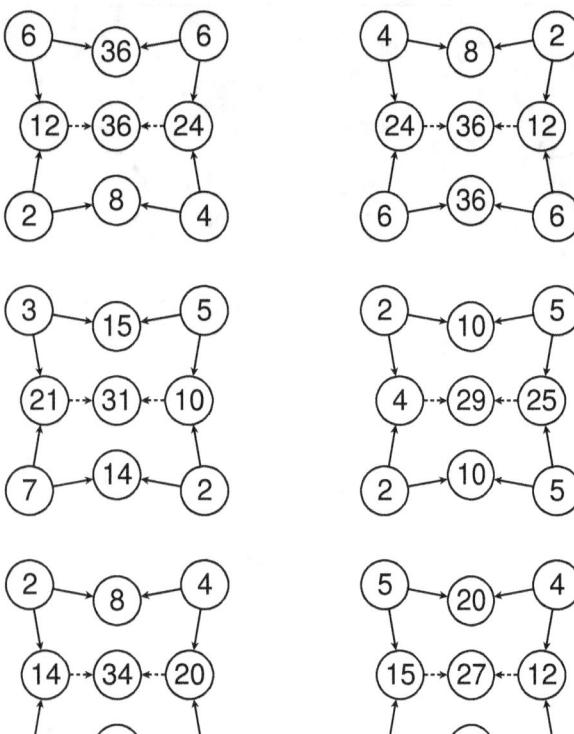

Solutions for Page 86

Find the missing numbers. Solid lines mean multiply.
Dotted lines mean add.

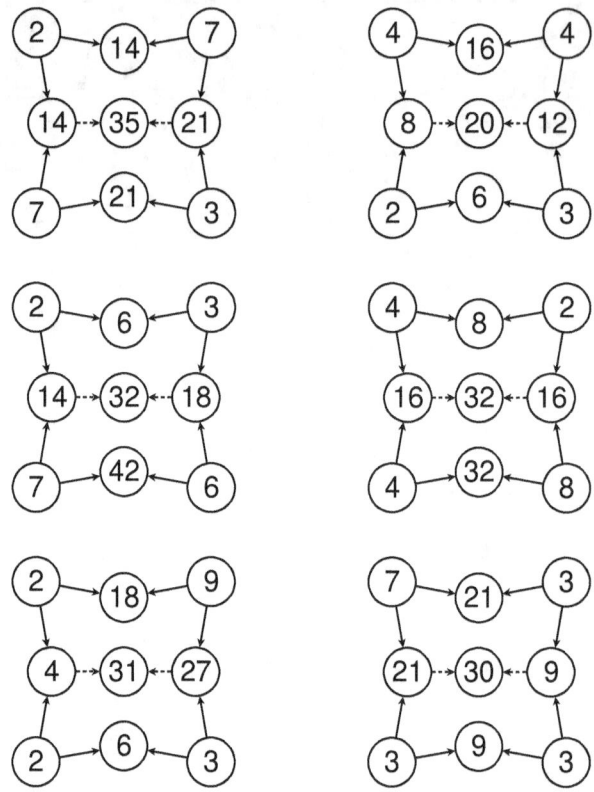

Solutions for Page 87

Find the missing numbers. Solid lines mean multiply.
Dotted lines mean add.

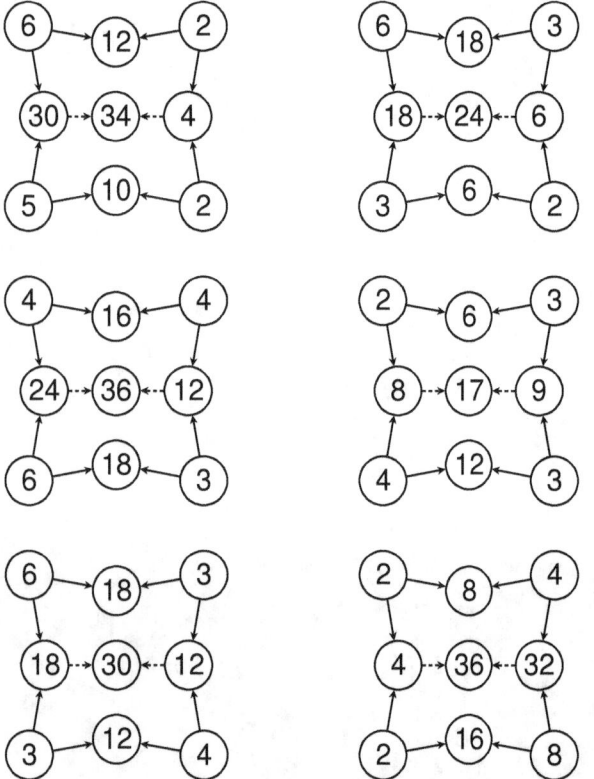

Solutions for Page 88

Find the missing numbers. Solid lines mean multiply.
Dotted lines mean add.

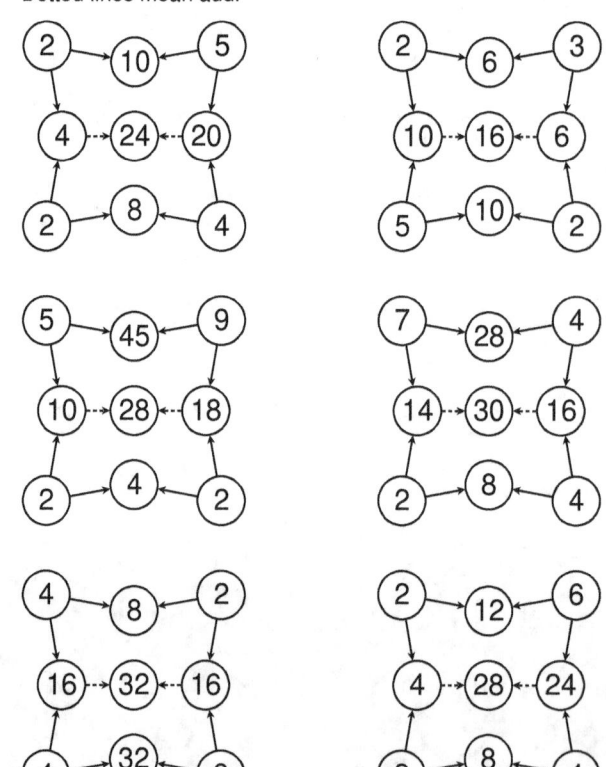

Solutions for Page 89

Find the missing numbers. Solid lines mean multiply.
Dotted lines mean add.

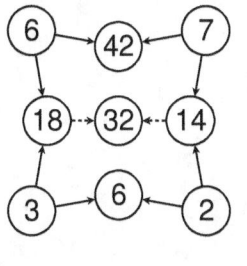 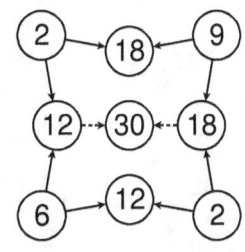

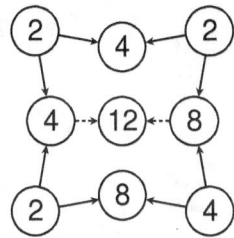 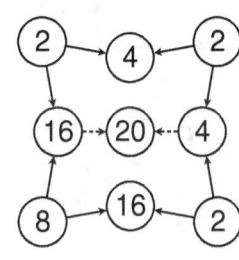

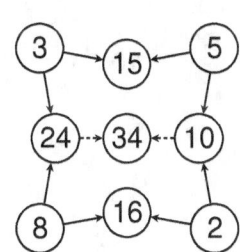 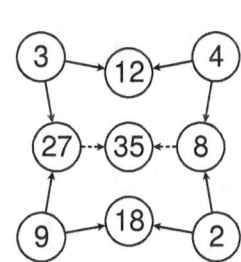

Solutions for Page 90

Find the missing numbers. Solid lines mean multiply.
Dotted lines mean add.

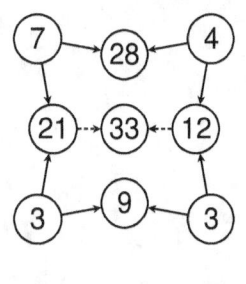 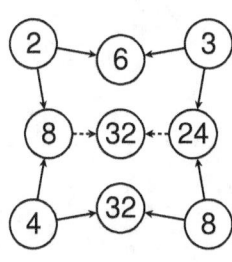

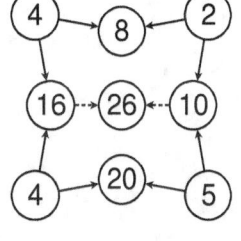 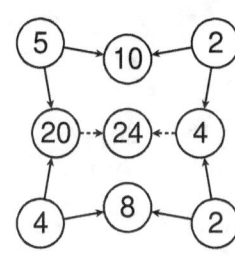

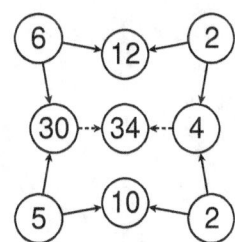 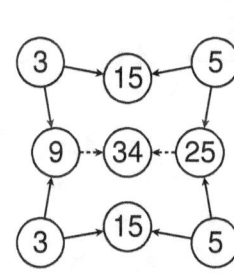

Solutions for Page 91

Find the missing numbers. Solid lines mean multiply.
Dotted lines mean add.

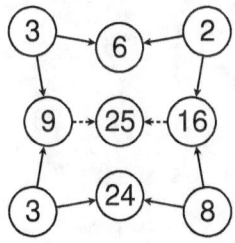 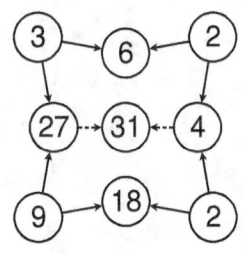

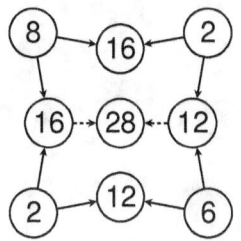 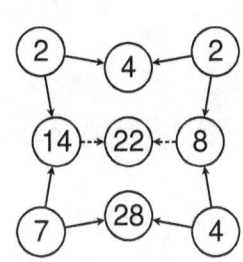

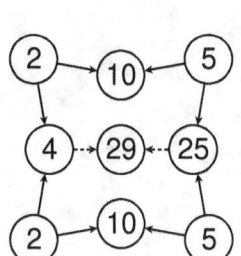 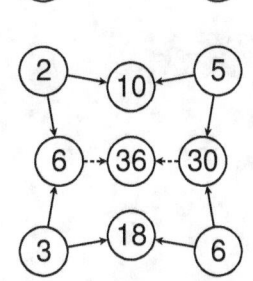

Solutions for Page 92

Find the missing numbers. Solid lines mean multiply.
Dotted lines mean add.

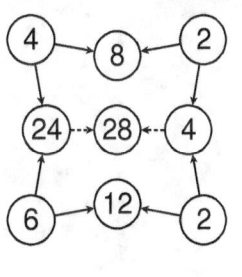 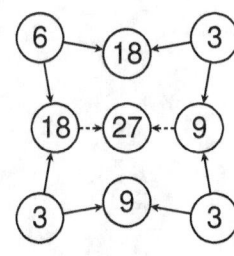

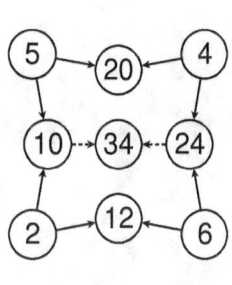 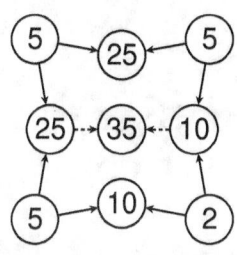

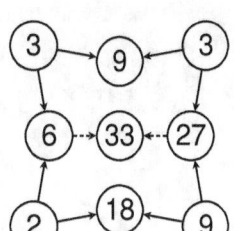 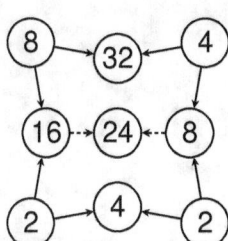

Solutions for Page 93

Find the missing numbers. Solid lines mean multiply.
Dotted lines mean add.

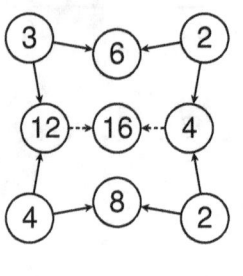
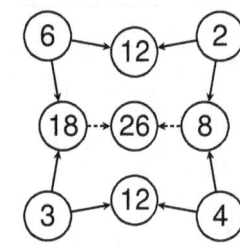

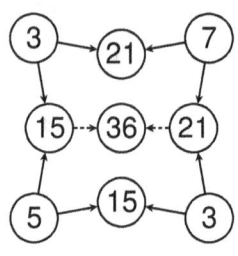
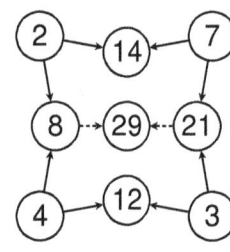

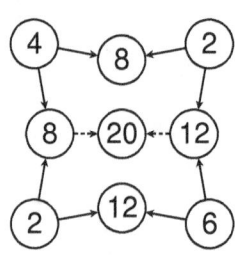
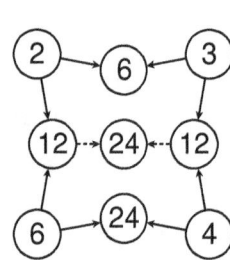

Solutions for Page 94

Find the missing numbers. Solid lines mean multiply.
Dotted lines mean add.

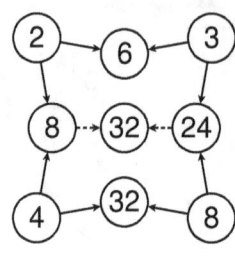
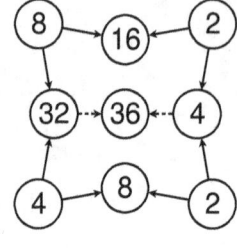

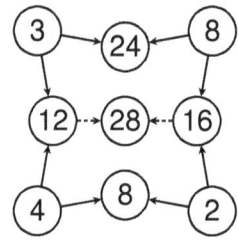
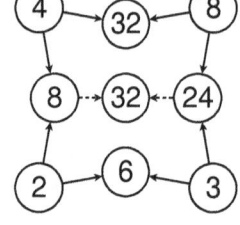

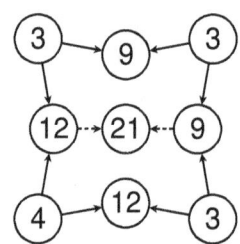
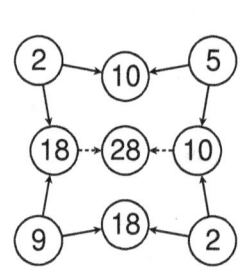

Solutions for Page 95

Find the missing numbers. Solid lines mean multiply.
Dotted lines mean add.

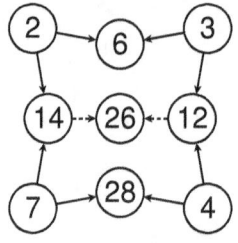
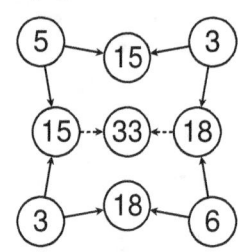

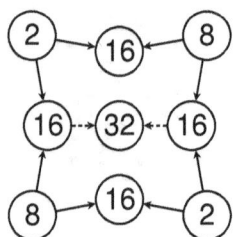
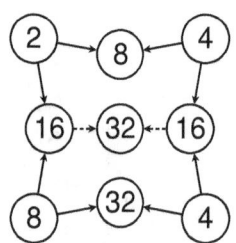

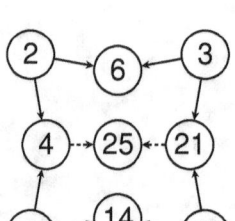
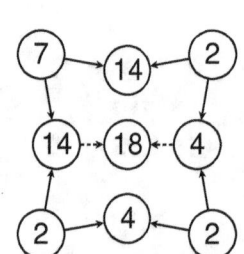

Solutions for Page 96

Find the missing numbers. Solid lines mean multiply.
Dotted lines mean add.

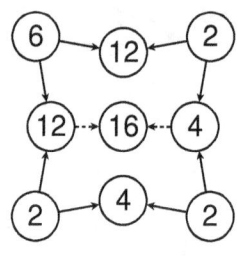
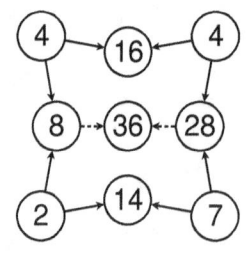

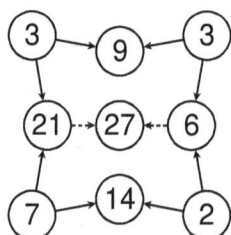
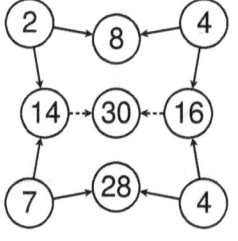

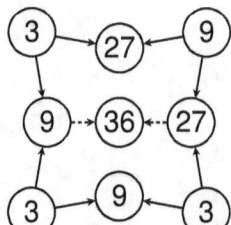
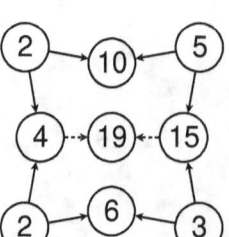

Solutions for Page 97

Find the missing numbers. Solid lines mean multiply.
Dotted lines mean add.

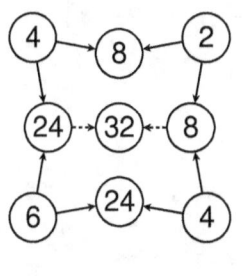 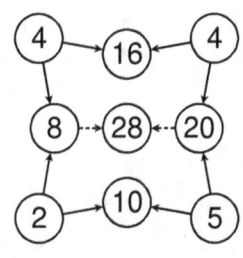

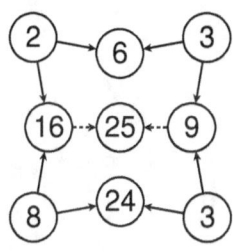 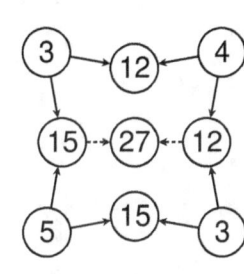

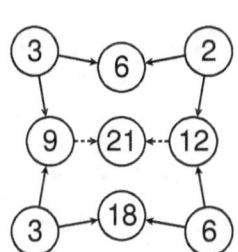 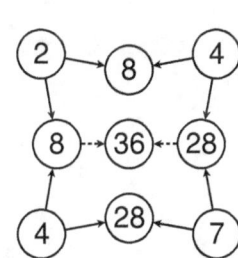

Solutions for Page 98

Find the missing numbers. Solid lines mean multiply.
Dotted lines mean add.

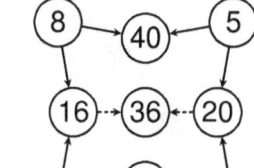 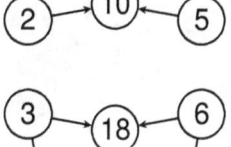

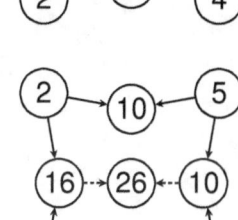 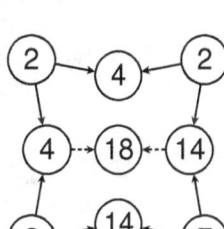

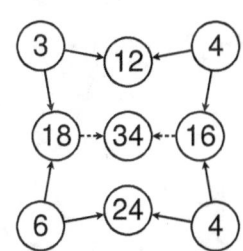

Solutions for Page 99

Find the missing numbers. Solid lines mean multiply.
Dotted lines mean add.

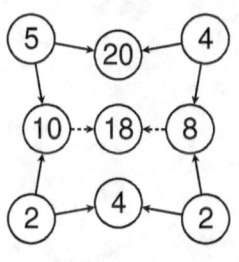 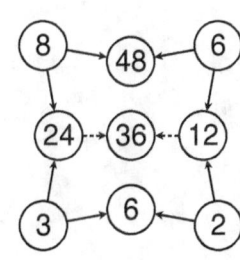

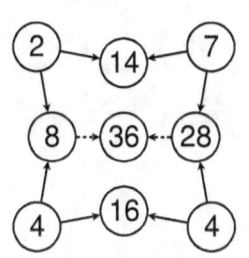 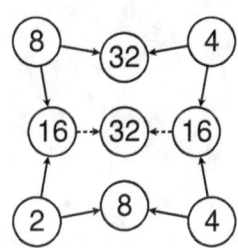

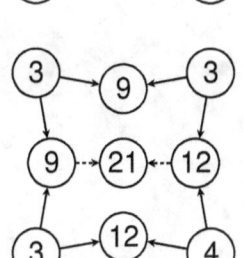 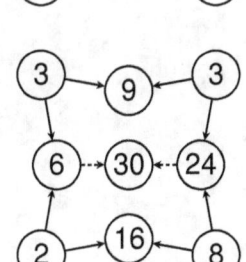

Solutions for Page 100

Find the missing numbers. Solid lines mean multiply.
Dotted lines mean add.

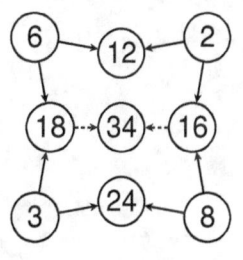 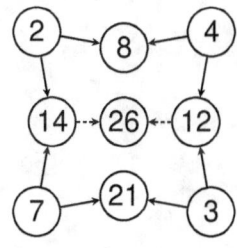

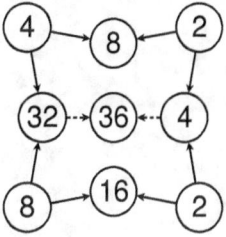 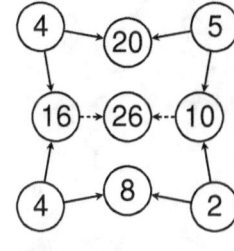

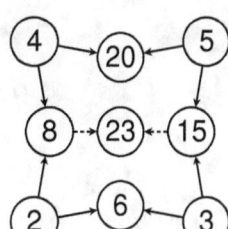 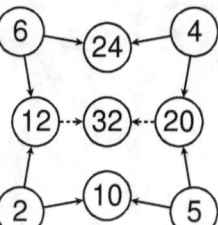